Principles of Data
Conversion System Design

Principles of Data Conversion System Design

Behzad Razavi

AT&T Bell Laboratories

IEEE PRESS

IEEE Circuits and Systems Society, *Sponsor*

The Institute of Electrical and Electronics Engineers, Inc., New York

This book may be purchased at a discount from the publisher when ordered in bulk quantities. For more information contact:

IEEE PRESS Marketing
Attn: Special Sales
P.O. Box 1331
445 Hoes Lane
Piscataway, NJ 08855-1331
Fax: (908) 981-8062

Printed in the United States of America
10   9   8   7   6   5   4   3   2

ISBN 0-7803-1093-4

IEEE Order Number: PC4465

Library of Congress Cataloging-in-Publication Data
Razavi, Behzad.
    Principles of data conversion system design / Behzad Razavi.
      p. cm.
    Includes bibliographical references and index.
    ISBN 0-7803-1093-4
    1. Analog-to-digital converters—design and construction.
    2. Digital-to-analog converters—Design and construction.
    3. Integrated circuits—Design and construction.   I. Title.
    TK7887.6.R39   1995
    621.39'814—dc20
                                     94-26694
                                          CIP

To the memory of my mother

Contents

Preface

Data conversion provides the link between the analog world and digital systems and is performed by means of sampling circuits, analog-to-digital (A/D) converters, and digital-to-analog (D/A) converters. With the increasing use of digital computing and signal processing in applications such as medical imaging, instrumentation, consumer electronics, and communications, the field of data conversion systems has rapidly expanded over the past twenty years. Monolithic integration, new architectures, and advances in integrated circuit (IC) technology have dramatically changed the design style of these systems and created new areas for research and development. As a result, the body of knowledge related to this field, primarily in the form of conference proceedings and journal papers, has grown to such extent that students and practicing engineers typically spend more than a year on the learning curve after they have completed other IC design courses. The lack of a systematic, comprehensive treatment of the subject has made the task of learning difficult and inefficient.

This book has been written as a unified text dealing with the analysis and design of data converters. Intended for classroom adoption as well as industrial practice, it methodically leads the reader from basic concepts to advanced topics while explaining design issues at both circuit and system level. In addition, to broaden the reader's view of technology-dependent design style, the text provides examples of CMOS, bipolar, and BiCMOS implementations for various circuits and discusses the trade-offs in each case.

The reader is assumed to have a solid understanding of analog IC design, preferably at the level of *Analysis and Design of Analog Integrated Circuits* by P. R. Gray and R. G. Meyer, and *Analog MOS Integrated Circuits for Signal Processing* by R. Gregorian and G. C. Temes. Some knowledge of digital circuits and the theory of signals and systems is also assumed.

The book consists of nine chapters. Chapter 1 serves as an introductory overview, familiarizing the reader with the role of data conversion in larger systems and providing the "big picture." Chapter 2 deals with basic sampling circuits and analyzes the behavior of MOS and bipolar switches with emphasis on their speed-precision trade-offs. Circuit techniques that relax such trade-offs are also described. Chapter 3 extends these techniques to the architecture level by introducing various sample-and-hold topologies.

Chapter 4 studies basic digital-to-analog conversion, viewing this function as reference multiplication or division. Topologies in which the reference is a voltage, current, or charge are analyzed and the switching functions required in such circuits are described. These concepts are applied to system-level design in Chapter 5, where digital-to-analog converter architectures are presented.

Chapter 6 deals with analog-to-digital converter architectures. Flash, two-step, interpolating, folding, pipelined, successive approximation, and interleaved architectures are studied and their design issues and sources of error are examined. Chapter 7 describes the design of building blocks of data conversion systems. Open-loop amplifiers, operational amplifiers, and comparators are discussed and means of improving their performance are introduced.

Chapter 8 focuses on precision techniques applicable to high-resolution data conversion. Comparator and op amp offset cancellation, D/A and A/D calibration, and overlap and digital correction are covered in this chapter.

Chapter 9 is concerned with the important topic of testing and characterization. Various approaches to evaluating the static and dynamic performance of sampling circuits and D/A and A/D converters are described in detail.

Each chapter is accompanied with an extensive set of references, allowing the reader to access the original work related to each topic, understand the intricate details in more depth, and learn techniques not described in the text.

Publishing a book is an elaborate, sometimes overwhelming task that can be carried out only with the support of a great many people. During the two years I worked on this book, the stimulating environment at AT&T Bell Labs and the guidance of my supervisor, Robert Swartz, enabled me to efficiently interleave research and writing. When the first draft was finished, a number of experts from both industry and academia reviewed various parts of the manuscript and provided helpful comments. In particular, Brian Brandt (IBM), Sing Chin (National Semiconductor), Robert Jewett (HP Labs), Andrew Karanicolas (AT&T Bell Labs), Stephen Lewis (UC Davis), Peter Lim (Chrontel), Krishnaswamy Nagaraj (AT&T Bell Labs), Marcel Pelgrom (Philips), David Rich (AT&T Bell Labs), and Bang-Sup Song (University of Illinois, Urbana-Champaign) contributed with their meticulous reviews, and I wish to express my gratitude to all of them. I am of course solely responsible for any errors or inconsistencies that may have remained in the text.

During the publication process, I have benefited from the kind support of the IEEE Press staff and would like to thank especially Russ Hall, Valerie Zaborski, Denise Gannon, and Dudley Kay for all their effort.

Behzad Razavi

1

Introduction to Data Conversion and Processing

The proliferation of digital computing and signal processing in electronic systems is often described as "the world is becoming more digital every day." Compared with their analog counterparts, digital circuits exhibit lower sensitivity to noise and more robustness to supply and process variations, allow easier design and test automation, and offer more extensive programmability. But, the primary factor that has made digital circuits and processors ubiquitous in all aspects of our lives is the boost in their performance as a result of advances in integrated circuit technologies. In particular, scaling properties of very large scale integration (VLSI) processes have allowed every new generation of digital circuits to attain higher speed, more functionality per chip, lower power dissipation, or lower cost. These trends have also been augmented by circuit and architecture innovations as well as improved analysis and synthesis computer-aided design (CAD) tools.

While the above merits of digital circuits provide a strong incentive to make the world digital, two aspects of our physical environment impede such globalization: (1) naturally occurring signals are analog, and (2) human beings perceive and retain information in analog form (at least on a macroscopic scale). Furthermore, when digital signals are corrupted by the medium such that they become comparable with noise, it is often necessary to treat them as analog signals. For example, according to information theory, for a digital signal buried in noise, amplitude digitization and subsequent decoding ("soft decision decoding") can improve the bit error rate.

In order to interface digital processors with the analog world, data acquisition and reconstruction circuits must be used: analog-to-digital converters (ADCs) to acquire and digitize the signal at the front end, and digital-to-analog converters (DACS) to reproduce the signal at the back end. This is illustrated in Figure 1.1.

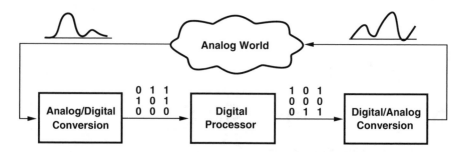

Fig. 1.1 Interface between analog world and a digital processor.

Data conversion interfaces find application in consumer products such as compact disc players, camera recorders (camcorders), telephones, modems, and high-definition television (HDTV), as well as in specialized systems such as medical imaging, speech processing, instrumentation, industrial control, and radar. We study one of these applications to illustrate the importance of both data conversion and digital processing in a typical product.

Figure 1.2 is a simplified block diagram of portable camcorder electronics [1]. The imaging front end consists of an array of charge-coupled devices (CCDs) that produce a charge output proportional to the light intensity. The charge packets from all the CCDs are sensed serially and converted to voltage, and the resulting signal is digitized by the ADC. Subsequently, operations such as autofocusing, image stabilization, luminance/chrominance (Y/C) processing, and zooming are performed using one or more digital signal processors (DSPs). The processed video signal is then converted to analog form and recorded on the tape.

While adding many features to the recorder and improving its user interface, the signal processing functions in Figure 1.2 are far too complex to be implemented in the analog domain. In fact, most of these functions have been added to camcorders simply because the ADC already provides the signals in digital form.

The performance required of the data conversion circuits used in video systems such as that of Figure 1.2 varies from one application to another. In portable camcorders, a conversion rate of a few tens of megahertz with 10-bit resolution is adequate, but the power dissipation (and preferably the

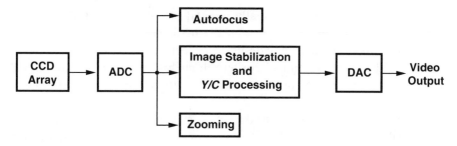

Fig. 1.2   Simplified block diagram of a portable camera recorder electronics.

supply voltage) must be minimized. In HDTV, speeds as high as 70 MHz are desirable, whereas in high-quality studio recording, resolutions of 12 to 14 bits are necessary.

Since data conversion interfaces must deal with both analog and digital signals, their design becomes increasingly difficult if they are to maintain comparable performance with their corresponding digital systems, i.e., not appear as a bottleneck in the signal path. This is because the primary trade-off in digital circuits is between speed and power, whereas that in analog circuits is between any two of speed, power, and precision (including res-olution, dynamic range, and linearity). Furthermore, the operation of both analog and digital circuits on the same chip leads to coupling of the noise generated by the digital section to the sensitive signals in the analog section. This coupling occurs via shared supply lines, substrate currents, or cross talk between adjacent lines.

High-performance data conversion systems have often been built as hy-brid structures, wherein different parts of the system are designed in different technologies and placed and interconnected on a common (nonconducting) substrate. This flexibility usually allows hybrids to achieve a higher speed than their monolithic counterparts—the key to their survival. However, issues such as cost, reliability, and power dissipation have created a trend toward im-plementing these interfaces in monolithic (VLSI) technologies and ultimately integrating an entire data processing system on a single chip. Most of the ar-chitectures and design concepts described in this book are used in both hybrid and monolithic applications, but the emphasis is on the latter type.

The integration of data conversion systems in VLSI technologies entails difficulties due to scaling, the very technique adopted to improve the per-formance of *digital circuits*. As supply voltages and device dimensions are reduced, many effects occur that are not predicted by the ideal scaling theory. For example, dynamic range becomes more limited, intrinsic gain of devices degrades, and device mismatch increases. In addition to these problems, many other analog design issues such as device noise and accurate control of device characteristics are usually ignored in optimizing VLSI technologies, and *modeling* of devices is typically performed with little concern for parameters important to analog design. Consequently, obtaining the required precision becomes the primary concern in analog and mixed analog-digital circuits, often necessitating conservative design and sacrifice in speed and power dissipation.

Let us now closely examine the data conversion interfaces of Figure 1.1. The analog-to-digital (A/D) interface converts a continuous-amplitude, continuous-time input to a discrete-amplitude, discrete-time signal. Shown in Figure 1.3 is this interface in more detail. First, an analog low-pass filter limits the input signal bandwidth so that subsequent sampling does not alias any unwanted noise or signal components into the actual signal band. Next, the filter output is sampled so as to produce a discrete-time signal. The amplitude of this waveform is then "quantized," i.e., approximated with a level from a set of fixed references, thus generating a discrete-amplitude signal. Finally, a digital representation of that level is established at the output.

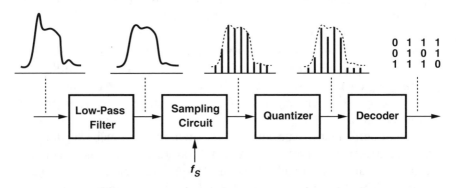

Fig. 1.3   Detailed analog-to-digital interface of Figure 1.1.

The ratio of the sampling rate f_S to the signal bandwidth distinguishes two classes of A/D converters. In "Nyquist-rate" ADCs, the sampling frequency is, in principle, slightly higher than twice the analog signal bandwidth to allow accurate reproduction of the original data. In "oversampling" converters, on the other hand, the signal is sampled at many times the Nyquist rate and subsequent digital filtering is utilized to remove the noise outside the signal bandwidth. These two classes require vastly different architectures and design techniques. In this book, we consider only Nyquist-rate converters. For oversampling data conversion, the reader is referred to the literature [2, 3].

The digital-to-analog (D/A) interface at the back end of the system shown in Figure 1.1 must convert a discrete-amplitude, discrete-time signal to a continuous-amplitude, continuous-time output. This interface is depicted in more detail in Figure 1.4. First, a D/A converter selects and produces an analog level from a set of fixed references according to the digital input. If the DAC generates large glitches during switching from one code to another, then a "deglitching" circuit (usually a sample-and-hold amplifier) follows to mask the glitches. Finally, since the reconstruction function performed by the DAC introduces sharp edges in the waveform as well as a sinc envelope in the frequency domain, an inverse-sinc filter and a low-pass filter are required to suppress these effects. Note that the deglitcher may be removed if the DAC is designed so as to have small glitches. Also, the inverse sinc filtering may be performed *before* D/A conversion, i.e., in the digital domain.

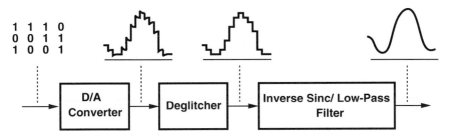

Fig. 1.4   Detailed digital-to-analog interface of Figure 1.1.

Figures 1.3 and 1.4 indicate that acquisition and reconstruction of data entail a great deal of mixed-signal processing: filtering, sampling, quantization, and digital encoding at the front end, and D/A conversion, sampling, and filtering at the back end. The design of data conversion interfaces demands a good understanding of various trade-offs in these operations as well as architecture and circuit techniques that improve the performance by relaxing these trade-offs.

In this book, we study sampling concepts and techniques in Chapters 2 and 3, D/A conversion in Chapters 4 and 5, and A/D conversion in Chapter 6. The important building blocks needed in performing these operations are described in Chapter 7, and methods of achieving high resolution in Chapter 8. Testing and characterization are the subject of Chapter 9.

REFERENCES

[1] A. Matsuzawa, "Low-Voltage and Low-Power Circuit Design for Mixed Analog/Digital Systems in Portable Equipment," *IEEE J. Solid-State Circuits*, vol. SC-29, pp. 470-480, April 1994.

[2] J. C. Candy and G. C. Temes, Editors, *Oversampling Delta-Sigma Data Converters*, IEEE Press, New York, 1992.

[3] S. Norsworthy, R. Schreier, and G. C. Temes, Editors, *Delta-Sigma Converters*, IEEE Press, New York, 1994.

2

Basic Sampling Circuits

Sampling circuits are often used at the front end of A/D converters to relax their timing requirements, and at the back end of D/A converters to suppress their glitch impulses. As such, they appear at the interface between the analog world and signal processing systems and must therefore achieve a precision and speed commensurate with the overall performance.

In this chapter, we describe basic circuits for analog signal sampling. Following a discussion of sampling techniques, performance metrics of sampling circuits are defined. Next, different types of sampling switches are studied and their parameters are compared. Finally, methods of improving the performance of MOS sampling devices are described.

2.1 GENERAL CONSIDERATIONS

A sampling circuit samples an analog signal and stores the result in a memory element until the next sampling instant. This operation is usually periodic and performed on voltages rather than currents because storing a voltage on a capacitor is easier than storing a current in an inductor.

Depending on the application, a given waveform can be sampled in different manners, resulting in different frequency spectra. Figure 2.1 illustrates three sampling techniques, called here ideal sampling [Figure 2.1(a)], zero-order hold [Figure 2.1(b)], and track and hold [Figure 2.1(c)]. In the first scheme, the signal $x(t)$ is multiplied by a periodic train of impulses:

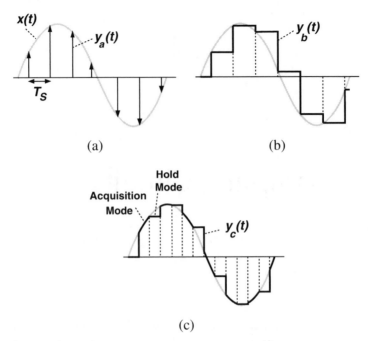

Fig. 2.1 Sampling schemes. (a) Ideal; (b) zero-order hold; (c) track and hold.

$$y_a(t) = x(t) \cdot \sum_{k=-\infty}^{+\infty} \delta(t - kT_S), \qquad (2.1)$$

where $\delta(\cdot)$ denotes the Dirac delta function. This causes the signal spectrum to be convolved with a train of impulses in the frequency domain, thus replicating and shifting the signal spectrum by integer multiples of T_S^{-1}:

$$Y_a(f) = X(f) * \sum_{n=-\infty}^{+\infty} \frac{1}{T_S}\delta(f - \frac{n}{T_S}) \qquad (2.2)$$

$$= \frac{1}{T_S} \sum_{n=-\infty}^{+\infty} X(f - \frac{n}{T_S}). \qquad (2.3)$$

This method yields signals that are easy to analyze but is not practical because of the difficulties in generating an ideal impulse or any reasonable approximation thereof. Also, the samples produced by this operation are often difficult to process because circuits following the sampler usually require that the sampled signal have a nonzero duration.

In the second scheme [Figure 2.1(b)], the value of the waveform at the sampling instant is captured and held until the next sampling instant. This is equivalent to multiplying the waveform by a periodic train of impulses and convolving the result with a rectangle function:

$$y_b(t) = \left[x(t) \cdot \sum_{k=-\infty}^{+\infty} \delta(t - kT_S) \right] * \Pi(\frac{t}{T_S} - \frac{1}{2}), \tag{2.4}$$

where $\Pi(t/T_S - 1/2)$ denotes a single pulse with unity amplitude from $t = 0$ to $t = T_S$. In the frequency domain, a spectrum similar to that of Figure 2.1(a) is obtained but it is multiplied by a sinc function:

$$Y_b(f) = e^{-j\pi f T_S} \frac{\sin \pi f T_S}{\pi f T_S} \sum_{n=-\infty}^{+\infty} X(f - \frac{n}{T_S}). \tag{2.5}$$

In practice, the input cannot be captured in zero time, and this method requires a sufficiently narrow sampling window (called the "aperture window") so as to provide an adequate approximation of the ideal zero-order hold.

In the third scheme [Figure 2.1(c)], the output tracks the input during sampling (usually called the "acquisition" or "tracking" mode) and remains at the last value of the input when the circuit enters the hold mode. Assuming the acquisition and hold modes each last $T_S/2$ seconds, we can decompose the waveform in Figure 2.1(c) into two modified square waves with a period of T_S: one whose amplitude is multiplied by the input signal, and another whose amplitude in each period is equal to the held value at the output:

$$y_c(t) = y_1(t) + y_2(t), \tag{2.6}$$

where

$$y_1(t) = x(t) \left[\Pi(\frac{2t}{T_S} - \frac{1}{2}) * \sum_{k=-\infty}^{+\infty} \delta(t - kT_S) \right] \tag{2.7}$$

$$y_2(t) = \left[x(t) \sum_{k=-\infty}^{+\infty} \delta(t - kT_S - \frac{T_S}{2}) \right] * \Pi(\frac{2t}{T_S} - \frac{1}{2}). \tag{2.8}$$

Thus,

$$Y_c(f) = \sum_{n=-\infty}^{+\infty} e^{-jn\pi/2} \frac{\sin(n\pi/2)}{n\pi} X(f - \frac{n}{T_S})$$

$$+ e^{-3j\pi f T_S/2} \frac{\sin(\pi f T_S/2)}{\pi f T_S} \sum_{n=-\infty}^{+\infty} X(f - \frac{n}{T_S}). \tag{2.9}$$

In contrast with the ideal sampler, the zero-order-hold and the track-and-hold schemes yield output spectra that have a sinc envelope, i.e., are "distorted." This problem mandates sinc compensation techniques [1] if the sampler is used at the back end of a data processing system, for example, following a D/A converter (Chapter 1). However, the zero-order-hold and track-and-hold circuits can be used at the *front end* of A/D converters with no concern for the sinc distortion. This is possible because A/D converters sense the output of the front-end sampler only during the hold mode, and hence the digitized value corresponds to sampled points on the input waveform. In other words, the combination of the front-end track-and-hold and the A/D converter operates as an ideal sampling circuit.

In high-speed systems, the distinction between the outputs of the zero-order-hold and track-and-hold schemes begins to diminish because the aperture window width becomes comparable with the sampling period. As a result, except for special applications where the aperture window is in the picosecond range [2], most monolithic sampling circuits operate as in the track-and-hold scheme. In this book, we consider implementations of this scheme, which are often called sample-and-hold amplifiers (SHAs) or track-and-hold amplifiers (THAs).

Figure 2.2 shows a simple sample-and-hold circuit. In the sampling (acquisition) mode, switch S (controlled by CK) is on and the output voltage, V_{out}, tracks the input voltage, V_{in}. In the transition to the hold mode, S turns off and V_{out} remains constant until the next sampling period. In this circuit, the switching operation and the transient currents drawn by C_H introduce noise at the input, often mandating the use of a front-end buffer. Furthermore, since the voltage stored on C_H during the hold mode can be corrupted by any constant or transient current drawn by the following circuit, a buffer must also be placed at the output, resulting in the circuit shown in Figure 2.3.

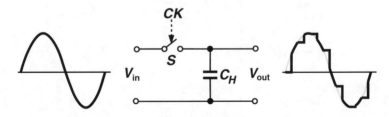

Fig. 2.2   Simple sample-and-hold circuit.

In practice, the nonidealities associated with the buffers and the sampling switch in Figure 2.3 necessitate substantial added complexity to achieve a given set of performance specifications. In fact, as discussed in Chapter 3, some SHA architectures are considerably different from that of Figure 2.3.

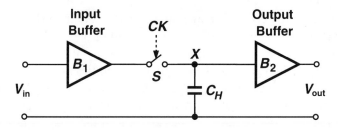

Fig. 2.3   Sample-and-hold circuit with input and output buffers.

Before describing various nonidealities that accompany the building blocks of SHAs, we need to define performance specifications of sampling circuits.

2.2 PERFORMANCE METRICS

In order to characterize sampling circuits thoroughly, a large number of parameters must be evaluated. The terminology and definitions adopted for SHA metrics by different manufacturers are not exactly the same and can cause confusion when different designs are compared. In this section, we define a number of terms commonly used to describe the performance of SHAs so as to establish a consistent set of metrics for this book. For a more comprehensive set of definitions, the reader is referred to the literature [3] and to manufacturers' data books.

The performance metrics defined below are illustrated graphically in Figure 2.4 and discussed using the SHA architecture of Figure 2.3.

- Acquisition time, t_{acq}, is the time after the sampling command required for the SHA output to experience a full-scale transition and settle within a specified error band around its final value. Acquisition time is determined by the recovery delay of B_1 and B_2, the on-resistance of S, the value of C_H, and the maximum allowable error.

- Hold settling time, t_{hs}, is the time after the hold command required for the SHA output to settle within a specified error band around its final value. This time is given primarily by the settling time of B_2.

- Dynamic range is the ratio of the maximum allowable input swing and the minimum input level that can be sampled with specified accuracy.

Dynamic range is limited by supply voltage, threshold or turn-on voltage of devices used in the circuit, and input-referred noise of the circuit.

- Nonlinearity error is the maximum deviation of the SHA input/output characteristic from a straight line passed through its end points [line AB in Figure 2.4(b)]. Usually specified for the *held* values of the output, this error originates from nonlinearities in B_1 and B_2, nonlinear dependence upon V_{in} of the charge injected by S onto C_H, and variation of the switch on-resistance with the input voltage.

- Aperture jitter is the random variation in the time required for the sampling switch to turn off after the hold command is asserted. Also called "aperture uncertainty," this error is a measure of the deviation of sampling instants from equally spaced points in time and arises from the noise that affects the hold command assertion (e.g., jitter in transitions of CK).

- Pedestal error voltage is the error introduced at the SHA output during the transition from sample to hold. This error stems from the charge injected by S onto C_H when this switch turns off.

- Gain error is the deviation of the slope of line AB from its ideal value (usually unity). This error results from the gain error of B_1 and B_2 and input-dependent pedestal voltage.

- Hold-mode feedthrough is the percentage of the input signal that appears at the output during the hold mode. This effect appears because switch S usually has a parasitic capacitive path between its input and output terminals even in the off state. This path conducts voltage variations and gives rise to input feedthrough during the hold mode.

- Droop rate is the rate of discharge of the capacitor during the hold mode. Droop rate is a function of the leakage currents drawn by parasitic dc paths from node X to other nodes (e.g., the substrate), the input bias current of B_2, and the value of C_H.

- Signal-to-noise ratio (SNR) is the ratio of the signal power to the noise power at the output in the hold mode (usually for a sinusoidal input). SNR is limited by the noise contributions from B_1, B_2, and S and the aperture jitter.

- Signal-to-(noise + distortion) ratio (SNDR) is the ratio of the signal power to the total noise and harmonic power at the output in the hold mode (for a sinusoidal input). SNDR is limited by the noise sources mentioned above and nonlinearities resulting from B_1, B_2, and charge injection of S.

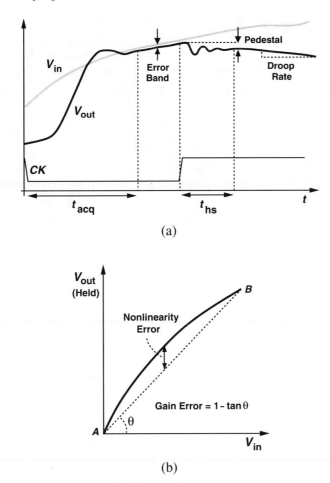

Fig. 2.4   Sample-and-hold performance metrics.

2.3 SAMPLING SWITCHES

As noted in the previous section, a large number of SHA limitations originate from nonidealities of the sampling switch. Acquisition time, aperture jitter, nonlinearity, pedestal error, feedthrough, and SNDR of these circuits are strongly influenced by the sampling switch performance.

In this section, we describe two types of sampling switches commonly utilized in CMOS and bipolar SHAs. In Chapter 3, we will see that, depending on the architecture, other switching techniques can be employed to improve the performance.

2.3.1 MOS Switches

An MOS transistor can be used as an analog switch, with its gate voltage controlling the resistance between its source and drain (Figure 2.5). For a square-law NMOS device that operates in the linear (triode) region, this resistance can be expressed as

$$R_{\mathrm{on}} = \frac{1}{\mu_n C_{\mathrm{ox}} \frac{W}{L}(V_{\mathrm{GS}} - V_{\mathrm{TH}})}, \tag{2.10}$$

where μ_n is the electron mobility in the channel, C_{ox} is the gate oxide capacitance per unit area, W and L are the effective width and length of the device, respectively, V_{GS} is the gate-source voltage, and V_{TH} is the threshold voltage. For a fixed sampling capacitor, the acquisition time can be decreased only by lowering R_{on}, i.e., by increasing the terms in the denominator of (2.10). In a given CMOS process, $\mu_n C_{\mathrm{ox}}$ is normally constant and V_{GS} usually cannot exceed the supply voltage, leaving W/L as the only variable in (2.10). Thus, high-speed applications often incorporate MOS switches with a large W/L.

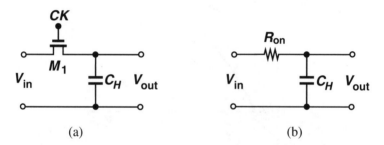

(a)   (b)

Fig. 2.5 (a) MOS sampling circuit; (b) equivalent circuit in the sampling mode.

In addition to a finite on-resistance, MOS switches exhibit channel charge injection and clock feedthrough. When on, a MOSFET carries a certain amount of charge in its channel that, under strong inversion conditions, can be expressed as

$$Q_{\mathrm{ch}} \approx W L C_{\mathrm{ox}}(V_{\mathrm{GS}} - V_{\mathrm{TH}}). \tag{2.11}$$

When the device turns off, this charge leaves the channel through the source and drain terminals, introducing an error voltage on the sampling capacitor (Figure 2.6). This error appears as an offset if Q_{ch} is constant, a gain error if Q_{ch} is linearly proportional to the input signal, or a nonlinear term if Q_{ch} has a nonlinear dependence on the input signal. While the first two types of error can be tolerated in some applications (e.g., data converters used in digital signal processing), the third type limits the linearity of the SHA and contributes harmonic distortion. The nonlinear component in Q_{ch} arises primarily from the nonlinear dependence of V_{TH} in (2.11) on the input voltage through body effect.

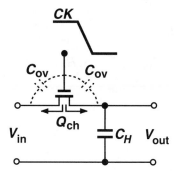

Fig. 2.6 MOS switch charge injection.

The charge injection mechanism in MOS switches has been analyzed extensively [4, 5, 6]. These studies show that the fraction of the charge injected onto the source and drain terminals depends on both the impedance seen at these nodes and the clock transition time [6]. In addition, these studies have provided mathematical descriptions of the injection mechanism and theoretical and experimental plots of the injected charge as a function of the impedances and clock transition time [6]. In practice, however, it is difficult to accurately predict or control these variables or apply the error figures measured for a given topology to another circuit. More importantly, most of the present circuit simulation programs do not model this mechanism accurately. For these reasons, many circuit techniques have been invented to suppress charge injection errors regardless of the exact value of such parameters as terminal impedances and clock transition times. These techniques are described in the context of SHA architectures in Chapter 3.

Another source of error in MOS switches is clock feedthrough, caused by the finite overlap capacitance between the gate and source or drain terminals. As depicted in Figure 2.6, when the gate control voltage CK changes state to turn off the switch, C_{ov} conducts the transition and changes the voltage stored on C_H by an amount equal to

$$\Delta V = \frac{C_{ov}}{C_{ov} + C_H} V_{CK}, \qquad (2.12)$$

where V_{CK} is the amplitude of CK. This equation indicates that clock feedthrough is independent of the input signal if C_{ov} is constant and thus appears as an offset in the input/output characteristic.

A frequency-dependent nonlinearity error in MOS sampling circuits arises from the variation of the switch on-resistance with the input voltage, i.e., the dependence of R_{on} in (2.10) upon V_{GS}. As shown in Figure 2.7, for high-frequency inputs this variation introduces input-dependent phase shift and hence harmonic distortion.

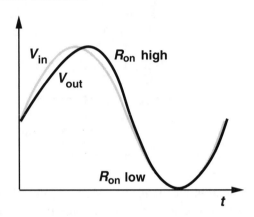

Fig. 2.7   Distortion caused by switch on-resistance variation in the tracking mode.

Another error that appears in high-speed MOS sampling circuits stems from the input-dependent sampling instant. Since the MOS switch turns off only when its gate-source voltage has fallen below V_{TH}, the time at which the device turns off (and the circuit enters the hold mode) depends on the instantaneous level of the input. For example, if the switch is an NMOS transistor, then the circuit enters the hold mode slightly later when the input signal is near ground potential than when it is higher. Illustrated in Figure 2.8, this phenomenon introduces jitter and harmonic distortion and becomes noticeable when the clock transition time is comparable with the input signal slew rate. For a sinusoidal input with amplitude A and frequency f_{in}, it has been shown that this phenomenon limits the signal-to-distortion ratio (SDR) of the SHA to

$$\mathrm{SDR}_{max} = 20 \log_{10} \frac{V_{CK}}{A f_{in} t_F} - 4.0, \qquad (2.13)$$

where V_{CK} and t_F are the clock amplitude and falltime, respectively [7].

An important aspect of sampling circuits is the hold-mode feedthrough because it can contribute noise to the output. As shown in Figure 2.9, for a MOS switch this error results from the path through the source-gate and gate-drain overlap capacitance and can be expressed as

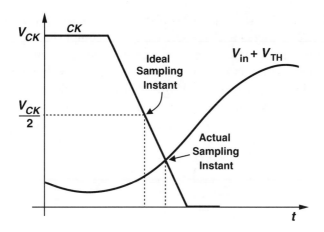

Fig. 2.8   Dependence of sampling instant on input level.

$$\frac{V_{\text{out}}}{V_{\text{in}}}\Big|_{\text{hold}} \approx \frac{C_{\text{ov}}}{C_H} \frac{R_{\text{out}}C_{\text{ov}}s}{2R_{\text{out}}C_{\text{ov}}s + 1}, \tag{2.14}$$

where R_{out} denotes the output resistance of the switch driver and it is assumed $C_{\text{ov}} \ll C_H$. The value of R_{out} should be chosen such that the feedthrough at maximum input frequency is sufficiently small.

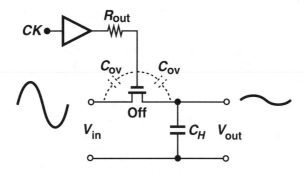

Fig. 2.9   Hold-mode feedthrough.

The sampling switch input and output range can also limit the full-scale voltage swing of a sampling circuit. For a supply voltage of V_{DD}, the circuit of Figure 2.5 has a maximum full-scale range of $V_{DD} - V_{\text{TH}}$, where V_{TH} includes the body effect where appropriate. In practice, the input swing hardly exceeds $(V_{DD} - V_{\text{TH}})/2$ because of the substantial increase in switch resistance and the resulting frequency-dependent harmonic distortion. This range can be extended to supply rails if the switch is realized as a complementary pair. As depicted in Figure 2.10, this is accomplished by controlling the gates of an NMOS and a PMOS device with complementary clocks so that the two devices turn on and off simultaneously. In this circuit, the NMOS transistor conducts for $0 \leq V_{in} < V_{DD} - V_{THN}$, while the PMOS device is on for $|V_{THP}| < V_{in} \leq V_{DD}$, thereby providing a rail-to-rail input and output range.

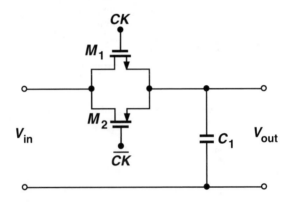

Fig. 2.10   Complementary MOS sampling switches.

In addition to an extended range, the circuit of Figure 2.10 has another important advantage over that of Figure 2.5: the equivalent on-resistance of a CMOS switch, compared with that of a single NMOS or PMOS device, varies much less as a function of the input voltage. Figure 2.11 plots the on-resistance of an NMOS, a PMOS, and a complementary MOS switch versus the input voltage, indicating only a small peak in the CMOS on-resistance near the middle of the range. The relatively constant on-resistance across the entire input/output range allows reasonable sizes for the switches and also minimizes the harmonic distortion caused by variation of switch resistance.

While it may seem that M_1 and M_2 in Figure 2.10 cancel each other's charge injection if they have identical dimensions, equation (2.11) indicates that $|Q_{ch}|$ depends on $|V_{GS} - V_{TH}|$ and hence will not be equal for M_1 and M_2 if V_{in} has an arbitrary value.

In high-speed applications, the complementary clocks required for CMOS switches can pose timing problems. Unless the clock edges are aligned such that the two transistors turn off at precisely the same instant, either one could be conducting weakly for a short while, introducing an input-dependent phase shift or sampling instant.

An important feature of MOS switches is that they introduce zero offset (level shift) from their source to their drain if the following circuit draws no

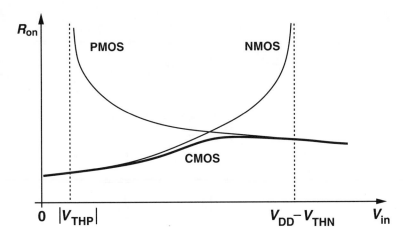

Fig. 2.11   Variation of on-resistance of NMOS, PMOS, and CMOS switches.

current. This is in contrast with the behavior of diode switches described below.

2.3.2 Diode Switches

Semiconductor diodes exhibit small on-resistance, large off-resistance, high-speed switching, and thus potential for the switching function in sampling circuits. A simplified diagram of a typical diode switch is shown in Figure 2.12 [8]. Here, four diodes form a bridge that provides a low-impedance path from V_{in} to V_{out} when current sources I_1 and I_2 are on and (in the ideal case) isolates V_{out} from V_{in} when I_1 and I_2 are off. Nominally, $I_1 = I_2 = I$.

For small signals at V_{in} and V_{out}, the equivalent on-resistance of the switch, R_{on}, is equal to the parallel combination of the resistance of the two branches consisting of D_1-D_2 and D_3-D_4. Each diode exhibits an incremental resistance, $1/g_m$, resulting from its exponential I-V characteristics, as well as an ohmic resistance, r_d, due to contact and material resistance. Thus,

$$R_{on} = \frac{1}{g_m} + r_d. \tag{2.15}$$

In order to minimize R_{on}, g_m can be increased by raising the bias current or r_d can be lowered by increasing the area of each diode. The latter remedy, however, increases the junction capacitance of the diodes, thus causing larger feedthrough in the hold mode. This effect is quantified later.

Actual implementations of the diode sampling bridge often control only *one* of the current sources, usually I_2, while the other is always on. This is because it is difficult to turn off I_1 and I_2 at precisely the same instant,

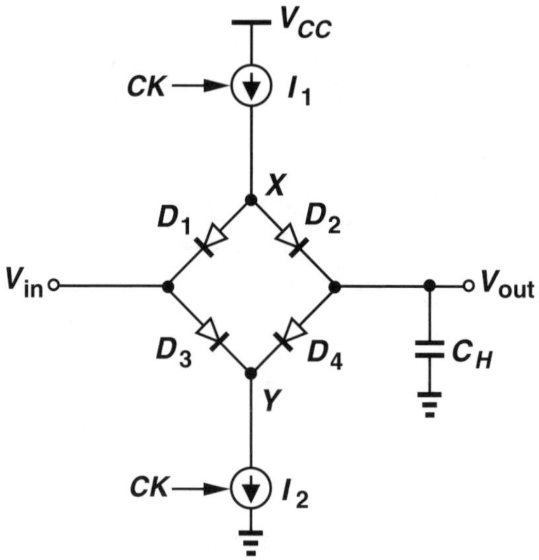

Fig. 2.12   Diode sampling switch.

especially if high-speed pnp or PMOS transistors are not available. Figure 2.13(a) illustrates such an implementation. Here, when CK is high, Q_1 is off, Q_2 is on, and the circuit is in the acquisition mode. When CK goes low, Q_1 turns on, Q_2 and hence the bridge turn off, and the circuit enters the hold mode. Note that I_1 need not be a high-speed device for it operates as a passive component, but the capacitance it introduces at node X is critical.

In the circuit of Figure 2.13(a), when CK is low and the bridge is off, Q_1 may enter saturation because the voltage at node X is not well-defined. Figure 2.13(b) depicts a modification where clamp diodes D_5 and D_6 and current source I_3 are added to the circuit. Typically, $I_2 \approx I_1 + I_3$ and the clamp voltage V_{B1} is at the midpoint of the input voltage range. When CK is high, Q_2 sinks both I_1 and I_3, and D_5 and D_6 are reverse-biased if $V_{B1} - V_{D1(on)} < V_{in} < V_{B1} + V_{D3(on)}$. When CK goes low, I_2 flows from node X, D_5 turns on, and V_X is clamped to $V_{B1} - V_{D5(on)}$, while I_3 flows through D_6 and V_Y is clamped to $V_{B1} + V_{D6(on)}$.

The variation of the on-resistance of a diode switch as a function of the input voltage is different from that of a MOS device. In the circuit of Figure 2.13, when the input voltage goes through a positive excursion, C_H draws current from D_2; therefore, D_1 and D_4 conduct less, whereas D_2 and D_3 conduct more. If the change in the currents of D_1-D_4 is a small fraction of their quiescent current I_D, then the increase in the resistance of D_1 and D_4 is compensated by the decrease in the resistance of D_2 and D_3, respectively,

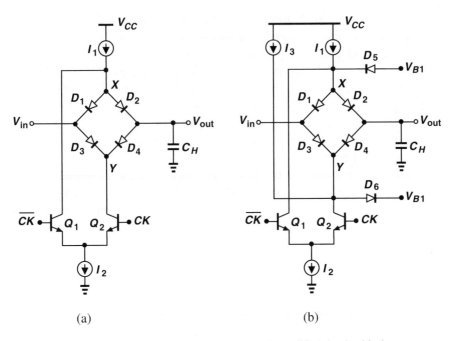

Fig. 2.13   (a) Simple bipolar sampling bridge; (b) modified circuit with clamp diodes.

thereby yielding negligible variation in the on-resistance of the overall switch. We can formulate the above condition by noting that if

$$V_{in} \approx V_{out} = A \sin \omega t, \qquad (2.16)$$

then the maximum current drawn by C_H occurs when $t = n\pi/\omega$ and is equal to

$$I_{max} = C_H \frac{dV_{out}}{dt}\Big|_{t=n\pi/\omega} \qquad (2.17)$$

$$= C_H A \omega. \qquad (2.18)$$

This current must remain much less than I; i.e.,

$$C_H A \omega \ll I. \qquad (2.19)$$

The charge injection and feedthrough behavior of diode switches is also different from that of MOS devices. Since a diode biased at a current I_D carries a charge equal to $I_D \cdot \tau_F$, where τ_F is the transit time, the charge injection error is relatively constant because, as explained above, in a properly designed switch I_D remains fairly constant. Thus, charge injection introduces only a constant offset in the input/output characteristic. On the other hand, in the diode bridge of Figure 2.13(b) the change in V_X and V_Y, which is coupled to the output through the junction capacitance of D_2 and D_4, causes nonlinearity and gain error [9]. To understand why, note that in the tracking mode, $V_X = V_{in} + V_{D(on)}$ and $V_Y = V_{in} - V_{D(on)}$, whereas in the hold mode, $V_X = V_{B1} - V_{D(on)}$ and $V_Y = V_{B1} + V_{D(on)}$. Consequently, in the transition from tracking to hold, V_X drops by $V_{in} - V_{B1} + 2V_{D(on)}$ and V_Y rises by $V_{B1} - V_{in} + 2V_{D(on)}$, creating an input-dependent pedestal at V_{out}. Since the junction capacitance of D_2 and D_4 is voltage-dependent, the pedestal has a significant nonlinear component.

In order to eliminate the input dependence of the pedestal, V_{B1} can be bootstrapped to the held output voltage [9]. This is illustrated in Figure 2.14, where the unity-gain buffer B_1 provides a clamp voltage equal to the held level. Now V_X and V_Y change by $-2V_{D(on)}$ and $+2V_{D(on)}$, respectively, yielding substantially less nonlinearity.

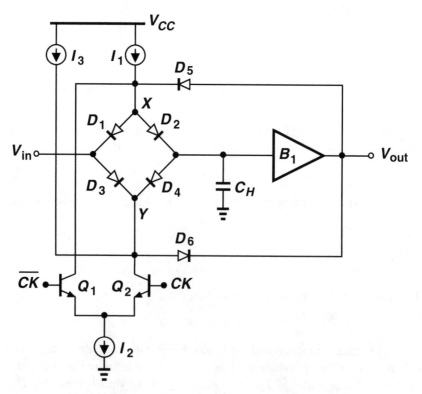

Fig. 2.14 Sampling bridge with bootstrapped clamp voltage.

The hold-mode feedthrough of sampling bridges depends on the junction capacitance of the bridge diodes and the on-resistance of the clamp devices.

For example, when the circuit of Figure 2.13(b) is in the hold mode, it can be simplified as shown in Figure 2.15, where the junction capacitance of D_1-D_4 is modeled with C_j, and the on-resistance of D_5 and D_6 with R_{on}. For $C_j \ll C_H$, the feedthrough transfer function is expressed as

$$\frac{V_{\text{out}}}{V_{\text{in}}}\Big|_{\text{hold}} \approx \frac{2C_j}{C_H} \frac{R_{\text{on}}C_j s}{2R_{\text{on}}C_j s + 1}. \tag{2.20}$$

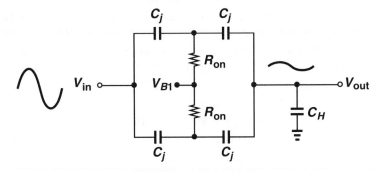

Fig. 2.15   Simplified circuit of diode bridge in the hold mode.

An important drawback of diode switches is the limitation they impose on the input and output voltage swings. In the circuit of Figure 2.13(a), current sources I_1 and I_2 typically require at least 0.5 V to maintain a high output impedance, the bridge consumes $2V_{D(\text{on})}(\approx 1.6 \text{ V})$ of the headroom, and Q_2 must have a V_{CE} of approximately 0.5 V to operate in the forward active region. As a result, the maximum voltage swing at V_{in} and V_{out} in a 5-V system is only 1.9 V.

Another drawback of diode switches is the dc offset from input to output due to device mismatches. For example, if in Figure 2.13(a), $I_1 = I_{C2} + \Delta I$ and D_3 and D_4 are mismatched, then the output offset voltage is

$$V_{\text{OS}} \approx V_T \ln(1 + \frac{\Delta I}{I_1} + \frac{\Delta I_S}{I_S}), \tag{2.21}$$

where $\Delta I_S / I_S$ is the relative mismatch between the saturation currents of D_3 and D_4.

2.3.3 Comparison of MOS and Diode Switches

Following our study of MOS and diode switches in the last two sections, we can now compare their properties.

- Diode switches generally have a lower on-resistance than MOS switches. If each diode in a bridge is biased at 0.5 mA and has a series (ohmic) resistance of 40 Ω, then the equivalent on-resistance of the bridge is approximately 90 Ω. Attaining such a low resistance with a MOSFET usually requires very large width-to-length ratios, typically greater than 1000. This in turn exacerbates charge injection and clock feedthrough problems.

- The on-resistance and charge injection of diode switches depend much less on the input voltage than do those of MOS devices, making the former more attractive for high-precision open-loop applications.

- Diode switches such as that of Figure 2.13 operate with clock voltage swings roughly an order of magnitude smaller than those of MOS circuits, allowing sharper edges and better definition of sampling points in time. For this reason and because of lower noise in ECL circuits than in MOS circuits, diode switches have a potentially lower jitter than their MOS counterparts.

- The input voltage range of MOS sampling circuits is generally larger than that of diode switches, thus allowing a wider dynamic range.

- MOS switches introduce no dc level shift (offset) from the input to the output if the following circuit draws no current. Diode bridges, on the other hand, suffer from a finite offset caused by mismatches in current sources and diodes.

- Diode switches are typically much more complex and dissipate much more power than MOS sampling circuits. The circuit of Figure 2.13(b), for example, requires at least six diodes, a differential pair, and three current sources whose magnitudes and variations with temperature and process must be well-controlled. A MOS sampling switch, on the other hand, consists of one or two transistors. While this difference in complexity and power dissipation may not be significant for a single SHA, it becomes important if a sampled-data system such as an A/D converter or a filter requires a great number of sampling switches.

2.3.4 Improvements in MOS Switch Performance

The simplicity of MOS switches has made them attractive for large-scale analog integrated circuits. However, as discussed in previous sections, MOS devices suffer from large on-resistance and substantial charge stored in their channel. In fact, the strong trade-off between these two parameters limits the level of speed-accuracy that can be achieved in a simple circuit such as that of Figure 2.5. We can formulate this trade-off by defining a figure of merit:

$$F = [(\text{pedestal error}) \cdot (\text{acquisition time})]^{-1} \qquad (2.22)$$

The pedestal error is

$$\Delta V_p = \frac{Q_{ch}}{2C_H} \tag{2.23}$$

$$= \frac{WLC_{ox}(V_{GS} - V_{TH})}{2C_H}, \tag{2.24}$$

where we have assumed half of the channel charge of the switch is injected onto C_H and neglected clock feedthrough. The acquisition time constant is

$$\tau_{acq} = R_{on}C_H \tag{2.25}$$

$$= \frac{C_H}{\mu_n C_{ox} \frac{W}{L}(V_{GS} - V_{TH})}. \tag{2.26}$$

Note that since τ_{acq} depends on the gate-source voltage and hence varies with input, modeling the acquisition behavior with a single time constant is only a rough approximation.

From (2.24) and (2.26), it follows that

$$F = \frac{1}{\Delta V_p \cdot \tau_{acq}} \tag{2.27}$$

$$= \frac{2\mu_n}{L^2}. \tag{2.28}$$

This equation indicates that, in a given CMOS technology, the MOS sampling circuit of Figure 2.5 does not achieve a speed-accuracy product higher than roughly $2\mu_n/L^2$. This product is further degraded by clock feedthrough.

In order to relax the trade-off given by (2.28), a number of circuit techniques have been proposed, two of which are illustrated in Figure 2.16. In Figure 2.16(a), a dummy device M_2 with half the width of the sampling switch M_1 (and the same length) is added and driven by $\overline{CK}$, the complement of the sampling clock CK [10]. In this circuit, when M_1 turns off and injects charge onto C_H, M_2 turns on and absorbs charge from C_H in its channel. Thus, if exactly half of the M_1 channel charge is injected onto C_H, then complete cancellation occurs and the held voltage on C_H is not corrupted by the charge injection. However, the fraction of channel charge injected by M_1 onto its source and drain depends on the impedance seen at the input and output nodes and the clock transition speed [6], indicating that C_H may not receive half of the M_1 channel charge and that this scheme may not provide accurate cancellation.

Figure 2.16(b) shows another sampling configuration, where the circuit is implemented in differential form [11]. In this topology, $V_{in,1}$ and $V_{in,2}$ are differential inputs (i.e., they vary by the same amount but in opposite

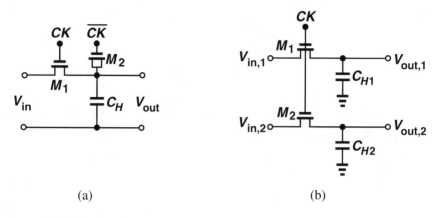

(a)   (b)

Fig. 2.16   Cancellation of MOS charge injection using (a) a dummy switch
and (b) differential operation.

directions) and $V_{out,1}$ and $V_{out,2}$ are differential outputs (i.e., their *difference*
is sensed by the following circuit). Charge injection and clock feedthrough
errors at $V_{out,1}$ and $V_{out,2}$ are to the first order equal and hence appear as a
common-mode (CM) component at the output. An important error, however,
still exists here: since the channel charge of M_1 and M_2 is a function of
their V_{GS} and since $V_{in,1} \neq V_{in,2}$, the differential held output includes an
input-dependent charge injection error term. This term introduces gain error
and nonlinearity, limiting the usefulness of this topology only to applications
where $V_{in,1} - V_{in,2}$ is much less than the $V_{GS} - V_{TH}$ of the switches.

Another MOS sampling technique is depicted in Figure 2.17. In the
acquisition mode, M_1 and M_2 are on, M_3 is off, $V_{out} = V_{DD}$, and the capacitor
voltage tracks the input. In the transition to the hold mode, first Φ goes low,
turning off M_2, and after a small delay Φ_Δ falls, turning M_1 off and M_3 on.
Thus, V_X drops from V_{in} to 0 and hence the *change* in V_{out} is equal to $-V_{in}$
at the sampling instant. Since M_2 always turns off first, the channel charge
of M_1—which is input-dependent—does not introduce any error. Moreover,
as the gate-source voltage of M_2 is independent of V_{in}, the channel charge
injected by this switch appears as a constant offset at the output.

While suppressing input-dependent charge injection, the circuit of Figure 2.17 suffers from another source of nonlinearity that limits its speed. In
the transition to hold, when V_{out} falls from V_{DD} to $V_{DD} - V_{in}$, a voltage division occurs between C_H and the drain junction capacitance of M_2, yielding
a nonlinear component in V_{out}. Since this capacitance is proportional to the
device width, it trades directly with the on-resistance and hence the acquisition time. Nevertheless, differential versions of this topology can provide
relatively high speed and high linearity.

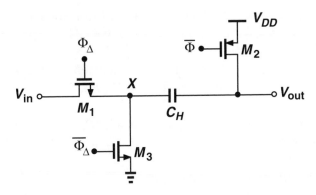

Fig. 2.17 CMOS sampling circuit with series capacitor.

Appendix 2.1 Effect of Aperture Jitter on SNR.   Consider a sine wave $V_{\text{in}} = A \sin 2\pi f_{\text{in}} t$ sampled at $t = kT_S + \epsilon$, where ϵ is the aperture jitter. Since each sampling instant can deviate from its ideal value by ϵ, the sampled amplitude has an error of $\epsilon\, dV_{\text{in}}/dt$. Thus, the overall sampled waveform can be viewed as the sum of an ideal sine wave and a noise component. To calculate the SNR for an otherwise ideal sampling circuit, we assume that ϵ is a random process uncorrelated with V_{in} and express the noise power as

$$P_\epsilon = \overline{\epsilon^2} \frac{1}{T_{\text{in}}^2} \int_0^{T_{\text{in}}} \left(\frac{dV_{\text{in}}}{dt}\right)^2 dt \tag{2.29}$$

$$= 2\pi^2 f_{\text{in}}^2 A^2 \overline{\epsilon^2}, \tag{2.30}$$

where $T_{\text{in}} = (2\pi f_{\text{in}})^{-1}$ and $\overline{\epsilon^2}(= \epsilon_{\text{rms}}^2)$ denotes the mean squared value of ϵ [12]. Thus,

$$\text{SNR} = -20\log(2\pi f_{\text{in}}\epsilon_{\text{rms}}) \quad \text{dB.} \tag{2.31}$$

This relation proves useful if jitter is the dominant source of noise in a system. In a general case, other sources of noise must be taken into account as well.

REFERENCES

[1] R. Gregorian and G. C. Temes, *Analog MOS Integrated Circuits for Signal Processing,* John Wiley and Sons, New York, 1986.

[2] K. Rush and D. J. Oldfield, "A Data Acquisition System for a 1-GHz Digitizing Oscilloscope," *Hewlett-Packard J.,* April 1986, pp. 4-11.

[3] S. K. Tewksbury, et al., "Terminology Related to the Performance of S/H, A/D, and D/A Circuits," *IEEE Trans. Circuits Syst.,* vol. CAS-25, pp. 419-426, July 1978.

[4] B. J. Sheu and C. Hu, "Switch-Induced Error Voltage on a Switched Capacitor," *IEEE J. Solid-State Circuits*, vol. SC-19, pp. 519-525, April 1984.

[5] W. B. Wilson et al., "Measurement and Modeling of Charge Feedthrough in N-Channel MOS Analog Switches," *IEEE J. Solid-State Circuits*, vol. SC-20, pp. 1206-1213, Dec. 1985.

[6] G. Wegmann, E. A. Vittoz, and F. Rahali, "Charge Injection in Analog MOS Switches," *IEEE J. Solid-State Circuits*, vol. SC-22, pp. 1091-1097, Dec. 1987.

[7] P. J. Lim, *Performance Limits of Circuits for Analog-to-Digital Conversion*, Ph.D. dissertation, Stanford University, March 1991.

[8] J. R. Gray and S. C. Kitsopoulos, "A Precision Sample and Hold Circuit with Subnanosecond Switching," *IEEE Trans. Circuit Theory*, vol. CT-11, pp. 389-396, Sept. 1964.

[9] K. Poulton, J. J. Corcoran, and T. Hornak, "A 1 GHz 6-Bit ADC System," *IEEE J. Solid-State Circuits*, vol. SC-22, pp. 962-970, Dec. 1987.

[10] C. Eichenberger and W. Guggenbuhl, "Dummy Transistor Compensation Of Analog MOS Switches," *IEEE J. Solid-State Circuits*, vol. SC-24, pp. 1143-1145, Aug. 1989.

[11] K. C. Hsieh et al., "A Low Noise Chopper-Stabilized Differential Switched Capacitor Filtering Technique," *IEEE J. Solid-State Circuits*, vol. SC-16, pp. 708-715, Dec. 1981.

[12] M. Shinagawa, Y. Akazawa, and T. Wakimoto, "Jitter Analysis of High-Speed Sampling Systems," *IEEE J. Solid-State Circuits*, vol. SC-25, pp. 220-224, Feb. 1990.

3

Sample-and-Hold Architectures

Since the introduction of the first monolithic sample-and-hold amplifier in 1974 [1], a variety of architectures amenable to integration in different technologies have been proposed. Owing to these architectures, as well as advances in integrated circuit technology, the performance of SHAs has dramatically improved, providing 12-bit acquisition times of less than 25 nsec in 1991 [2] compared to 10 μsec in 1974 [1]. These architectures generally employ circuit techniques to reduce the pedestal error without sacrificing speed and linearity of the system.

In this chapter, we describe a number of SHA architectures often used in data acquisition systems. Up to the mid-1980s, most sample-and-hold circuits fell into either the "open-loop" or the "closed-loop" category [1]. However, the configurations introduced in recent years cannot really be classified in this fashion because they incorporate various local and global feedback paths, obscuring the distinction between open-loop and closed-loop topologies in the conventional sense. For this reason, we describe each of these architectures individually.

3.1 CONVENTIONAL OPEN-LOOP ARCHITECTURE

The open-loop architecture has been considered attractive because of its simplicity and potential speed. Used in Chapter 2 to illustrate SHA properties and shown in Figure 3.1, this architecture consists of an input buffer B_1, a sampling circuit comprising S and C_H, and an output buffer B_2. When S is on, the circuit is in the acquisition mode and V_{out} tracks V_{in}. When S turns off, the instantaneous value of the input is stored on C_H and the circuit enters the hold mode.

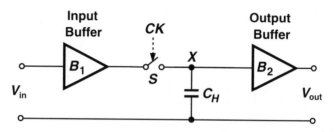

Fig. 3.1   Open-loop sample-and-hold architecture.

As this topology includes no global feedback, it is unconditionally stable (if B_1 and B_2 are stable) and can therefore be designed for high-speed operation. For example, utilizing Schottky diode bridges, this architecture has achieved rates as high as 2 GHz [3].

The speed of the circuit in Figure 3.1 is determined by its acquisition time and hold settling time. The acquisition time depends on the tracking speed and output impedance of B_1, the on-resistance of S, and the value of C_H, while the hold settling time is governed by the settling behavior of B_2.

In addition to their impact on the SHA's speed, B_1 and B_2 also influence the linearity of the system, especially because both experience full signal swings at their input and output. As a consequence, the linearity requirements of the overall SHA often impose restrictions on the design of B_1 and B_2, thus limiting their speed. In general, these buffers can employ open-loop configurations with various correction techniques [4] to achieve linearities up to 10 bits. For higher linearity, they are usually implemented as high-gain amplifiers with local feedback. Chapter 7 describes various amplifier design issues.

An important drawback of this architecture results from the input-dependent charge injected by the sampling switch onto the hold capacitor, an eminent source of nonlinearity in MOS implementations. As explained in Section 2.3, this error is not reliably canceled by dummy devices or differential configurations and hence limits the linearity of open-loop CMOS SHAs to approximately 8 bits. Diode switches, on the other hand, exhibit much less input-dependent charge injection and have been successfully used for linearities up to 12 bits in bipolar technology [5].

In summary, the open-loop architecture offers high speed and relatively high linearity when realized in bipolar SHAs with diode bridges but suffers from input-dependent pedestal error in CMOS implementations.

3.2 CONVENTIONAL CLOSED-LOOP ARCHITECTURE

In order to suppress input-dependent pedestal errors in a SHA, the sampling switch can be included in a feedback loop such that it experiences voltage swings much smaller than the input and output swings. This concept is the basis for the closed-loop architecture shown in Figure 3.2, which consists of a transconductance amplifier G_m, sampling devices S and C_H, and a voltage amplifier A_0 [1]. The circuit operates as follows. In the acquisition mode, S is on and the circuit functions as a two-stage op amp compensated by C_H and configured as a unity-gain buffer. Thus, the output closely follows the input and, if A_0 is large, X is a virtual ground node, allowing the voltage across C_H to track the input. When S turns off, the instantaneous output voltage is stored on C_H and the feedback circuit consisting of A_0 and C_H retains the sampled voltage at the output.

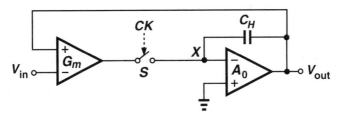

Fig. 3.2   Closed-loop sample-and-hold architecture.

An important feature of this architecture arises from the virtual ground property of node X: since in the sampling mode the output voltage of G_m is also close to ground potential, switch S always turns off with a constant voltage at its input and output terminals, thereby injecting a constant charge onto C_H and introducing a pedestal error that is independent of the input signal. As a result, this error appears primarily as an offset voltage and contributes negligible nonlinearity. In order to reduce the offset resulting from charge injection and clock feedthrough, a replica of the sampling network can be placed at the noninverting input of A_0 so that the pedestal appears equally at both of its inputs, i.e., as a common-mode voltage. This technique is illustrated for MOS switches in Figure 3.3 [6], where M_2 and C_2 are identical with M_1 and C_H, respectively. In this circuit, M_1 and M_2 turn off simultaneously, injecting channel charge onto C_H and C_2. However, these charge packets may not be

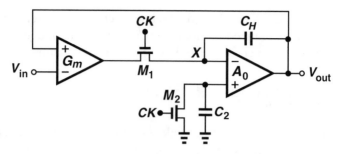

Fig. 3.3   Closed-loop sample-and-hold architecture with pedestal cancellation.

exactly equal because M_1 sees the output impedance of G_m on one side while M_2 sees the ground. Nonetheless, clock feedthrough components of the pedestals are equal and cancel out. More accurate cancellation of the pedestals can be achieved through the use of this architecture in fully differential form [6].

The main limitation of closed-loop architecture arises from its stability and speed considerations. Since the circuit of Figure 3.2 functions as a two-stage op amp in the sampling mode, the dominant pole given by the output impedance of G_m and the Miller multiplication of C_H must provide a reasonable phase margin so that the output quickly tracks the input with the required accuracy. As in a typical two-stage op amp (Chapter 7), several factors degrade the phase margin and, more importantly, the output settling behavior. Since G_m and A_0 usually contribute several nondominant poles, some of which may not be sufficiently greater than the dominant pole, settling to high accuracies may be slow. Furthermore, the pole given by the output impedance of A_0 and the load capacitance often causes long settling times. Additionally, the magnitude of this pole may *vary* with V_{out} if the output impedance of A_0 depends on the load current, thereby introducing output-dependent settling components [6]. This effect is discussed in Chapter 7.

The above stability issues often necessitate conservative compensation of the closed-loop architecture so as to avoid underdamped settling despite variations in process, load capacitance, and temperature. Consequently, this architecture does not usually achieve the maximum potential speed of a given technology.

Another drawback of this architecture stems from the signal path from V_{in} to V_{out} through the input capacitance of G_m. This path introduces significant hold-mode feedthrough if the capacitance between the input terminals of G_m is large, e.g., if the input stage of G_m utilizes large devices. The feedthrough is attenuated by the output impedance of the unity-gain amplifier comprising A_0 and C_H. But this impedance typically increases with the input frequency, allowing larger feedthrough of high-speed signals.

In summary, the closed-loop architecture suppresses the input-dependent component of the hold pedestal by incorporating the sampling switch in a feedback loop. This architecture, especially in a fully differential configuration, is attractive for high-precision systems but usually suffers from slow time response.

3.3 OPEN-LOOP ARCHITECTURE WITH MILLER CAPACITANCE

The conventional open-loop architecture described in Section 3.2 suffers from a fundamental limitation due to the speed-precision trade-off of the sampling switch given by (2.28). This trade-off results from the relationship between the on-resistance and channel charge of MOSFETs, indicating that the hold pedestal can be reduced only if slower acquisition is acceptable. However, noting that these limitations exist simply because the same capacitor is used for both acquisition and hold, we can avoid them by using *different* capacitors in the sampling and hold modes. The open-loop architecture with Miller capacitor is based on this concept and illustrated in Figure 3.4(a) [7].

The circuit consists of a sampling switch M_1 and an ac-coupled Miller amplifier comprising A_0, M_2, C_1 and C_2. In the sampling mode, both M_1 and M_2 are on, A_0 is configured as a unity-gain circuit, providing virtual ground at nodes X and Y, and capacitors C_1 and C_2 track the input voltage [Figure 3.4(b)]. In the transition to the hold mode, M_1 and M_2 turn off simultaneously and C_1, C_2, and A_0 form a feedback amplifier that introduces a capacitance of approximately $A_0 C_2$ from node Z to ground [Figure 3.4(c)]. Thus, in this architecture the hold capacitor is roughly $A_0 C_2/(C_1 + C_2)$ times the acquisition capacitor, thereby relaxing the speed-precision trade-off described in Section 2.3 by the same factor. Note that since M_2 always turns off with (virtual) ground potential at its source and drain, its charge injection causes negligible nonlinearity.

In this architecture, even though A_0 must be a high-speed amplifier to provide a low output impedance at high frequencies, it nonetheless does not need a wide dynamic range because its output voltage swing results from only the charge injected by M_2. This simplifies its design, allowing optimization for speed.

In practice, the topology of Figure 3.4(a) suffers from second-order sources of error. First, owing to their input-dependent switching point, M_1 and M_2 do not always turn off simultaneously, thus creating Miller effect either in the sampling mode (which slows down the acquisition and introduces input-dependent delay) or *after* M_1 has turned off and injected its charge onto C_1 and C_2 (in which case the amplifier will not suppress the error).

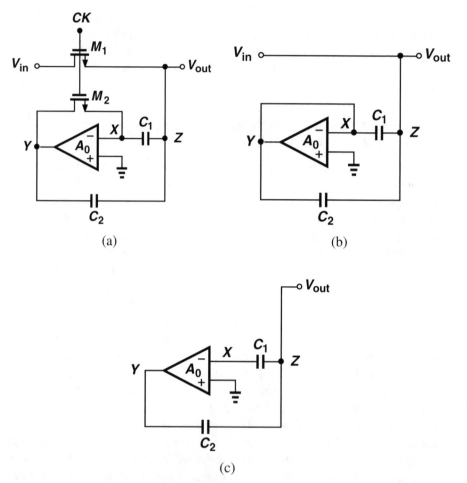

Fig. 3.4   Open-loop architecture with Miller capacitance. (a) Basic circuit;
(b) equivalent circuit in the acquisition mode; (c) equivalent circuit
in the hold mode.

Second, when turning off, M_1 and M_2 interact through C_1 and influence
each other's charge injection, making the charge injected by M_2 somewhat
input-dependent. Nonetheless, Lim and Wooley [7] have shown that the
nonlinearity introduced by this interaction is negligible for resolutions up to
8 bits.

   In summary, the open-loop architecture with Miller capacitance employs two different values of capacitance in the acquisition and hold modes
to achieve high speed and small pedestal error. This is accomplished using a
Miller amplifier that multiplies the effective value of the sampling capacitor
by a large number when the SHA enters the hold mode.

3.4 MULTIPLEXED-INPUT ARCHITECTURES

A class of SHA architectures employs input multiplexing to reconfigure the circuit when it goes from the acquisition to the hold mode. In this section, we describe two variants of this architecture.

Figure 3.5(a) shows the single-ended version of a multiplexed-input SHA originally proposed by Ryan [8] and later modified by Petschacher et al. [9]. It consists of transconductance amplifiers G_{m1} and G_{m2} and transresistance amplifier R. Nominally, $G_{m1}R = G_{m2}R = 1$. Amplifiers G_{m1} and G_{m2} are controlled (i.e., multiplexed) by CK and $\overline{CK}$. During sampling, G_{m1} is enabled, G_{m2} is disabled, and G_{m1} and R operate as a unity-gain amplifier, allowing V_{out} to track V_{in}. Note that the acquisition time constant is given primarily by the output resistance of R and the value of C_H. In the transition to the hold mode, G_{m1} is disabled, G_{m2} is enabled, and G_{m2} and R are configured as a unity-gain amplifier, thereby retaining the sampled value of V_{in} across C_H.

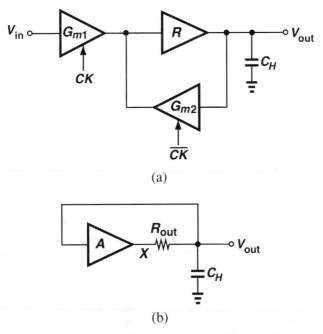

(a)

(b)

Fig. 3.5   Multiplexed-input architecture. (a) Basic (single-ended) circuit; (b) equivalent circuit in the hold mode.

In order to illustrate the hold-mode operation, we consider a simplified version of the circuit, shown in Figure 3.5(b), where $A = G_{m2}R$ (≈ 1) and R_{out} represents the open-loop output resistance of the amplifier. Assuming that C_H is charged to a voltage V_0 at the end of the acquisition mode and neglecting the input bias current of A, we note:

$$\frac{V_{\text{out}} - V_X}{R_{\text{out}}} = -C_H \frac{dV_{\text{out}}}{dt} \tag{3.1}$$

and

$$V_X = A V_{\text{out}}. \tag{3.2}$$

Thus,

$$V_{\text{out}} = V_0 \exp \frac{-t}{\tau}, \tag{3.3}$$

where $\tau = R_{\text{out}} C_H / (1 - A)$ and the origin of time is the beginning of the hold mode. Equation (3.3) shows that if $A = 1$, then $\tau = \infty$; i.e., the droop rate is zero and V_{out} will remain at V_0 indefinitely. If $A = 1 - \epsilon$, then V_{out} decays with a time constant equal to $R_{\text{out}} C_H / \epsilon$; i.e.; the droop time constant is $1/\epsilon$ times the acquisition time constant.

Several aspects of this architecture make it attractive for implementation in bipolar technology. First, G_{m1} and G_{m2} can be realized as simple differential pairs that are multiplexed by means of a third differential pair, yielding a fast sample-to-hold transition and low aperture jitter. Second, if G_{m1} and G_{m2} are identical, the charge injected onto the input node of R by G_{m1} at the end of the sampling mode is absorbed by G_{m2}, thus giving a small pedestal error. Third, the G_m and R stages can be implemented as low-gain, high-speed circuits because $G_{m1}R$ and $G_{m2}R$ need only be unity. This is particularly important if the process provides no vertical pnp transistors and hence prohibits the use of high-gain stages.

While achieving high speed, the architecture of Figure 3.5 presents several difficulties if employed for high-resolution applications. Since the acquisition time constant and the droop rate trade off according to the deviation of A from unity, G_{m1}, G_{m2}, and R often require correction techniques to reduce that deviation [9] and approach the desired combination of acquisition speed and droop rate. Note that since this correction should remain effective for the entire input range, the nonlinearity (i.e., variations in the gain) of the circuit must be as low as the gain error. These correction techniques, however, normally consume a substantial portion of the headroom, limiting the input voltage swing and the dynamic range. For example, $G_{m1}R$ can be implemented as shown in Figure 3.6 [9], where resistive degeneration in the emitters and diode compensation in the collectors of Q_1 and Q_2 are utilized to achieve a gain close to unity (Chapter 7). Considering the voltage drop across emitter and collector resistors and the headroom required for the clocked tail current source, we note that the input range hardly exceeds 1 V with $V_{CC} = 5$ V.

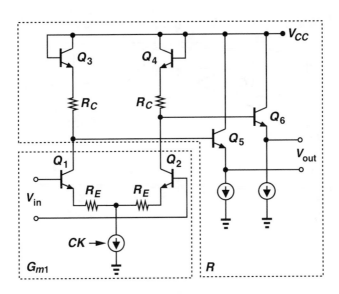

Fig. 3.6   Linearized amplifier used in open-loop SHAs.

Another input-multiplexed SHA architecture is shown in Figure 3.7 [10]. This topology consists of transconductance amplifiers G_{m1} and G_{m2} multiplexed by CK and $\overline{CK}$, transresistance amplifier R, and sampling devices S_1, S_2, C_1, and C_2. In the sampling mode, G_{m1} is enabled, G_{m2} is disabled, and S_1 and S_2 are on. Thus, the circuit is configured as shown in Figure 3.8(a), $G_{m1}R$ operates as a high-gain op amp, and the closed-loop gain is equal to $1 + R_2/R_1$. In other words, $V_{out} = (1 + R_2/R_1)V_{in}$.

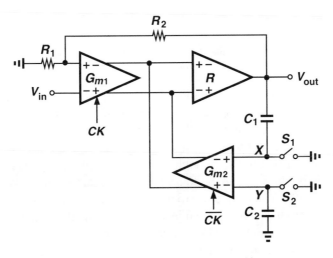

Fig. 3.7   Dual-loop multiplexed-input architecture.

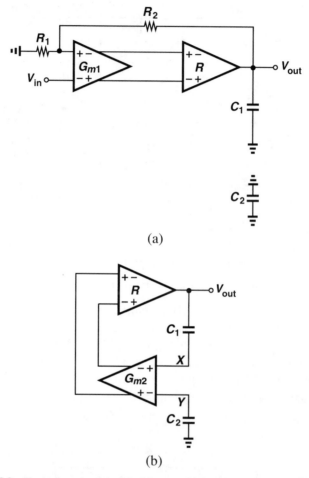

Fig. 3.8   Equivalent circuits of dual-loop multiplexed-input architecture. (a) Acquisition mode; (b) hold mode.

In the transition to the hold mode, G_{m1} is disabled, G_{m2} is enabled, and S_1 and S_2 turn off, yielding the hold configuration shown in Figure 3.8(b). Here, V_Y is close to zero and $G_{m2}R$ and C_1 function as a unity-gain amplifier, thereby maintaining an output voltage equal to that stored on C_1.

The principal feature of this architecture is its input-independent pedestal error. This is because S_1 and S_2 do not experience large voltage swings; i.e., they turn off with V_X and V_Y close to the ground potential. Consequently, the nonlinearity introduced by charge injection is quite small. Furthermore, if the sampling capacitors and switches are identical and the impedance in series with S_1 (i.e., the output impedance of $G_{m1}R$) is small, then the pedestals produced at the two inputs of G_{m2} are equal, yielding a zero pedestal at the output.

It is instructive to compare the two multiplexed-input architectures described in this section. While the topology shown in Figure 3.5 suffers from trade-off between acquisition speed and droop rate—necessitating a $G_{m2}R$ of precisely 1—the topology of Figure 3.7 has no such trade-off. Moreover, since the latter employs closed-loop amplifiers, it can achieve smaller nonlinearity, but at the cost of a potentially slower time response. On the other hand, both architectures have limited input range due to the stacked devices required for multiplexing.

3.5 RECYCLING ARCHITECTURE

The recycling architecture is another topology in which the sampling switch experiences small swings and hence introduces only a constant pedestal error. Shown in Figure 3.9 in simplified form, this architecture consists of two unity-gain buffers B_1 and B_2, a transconductance amplifier G_m, and a sampling circuit comprising S_1-S_5 and C_1-C_3 [11]. In the sampling mode, S_1-S_4 are on, S_5 is off, and the circuit is configured as shown in Figure 3.10(a). In this mode, G_m operates as a unity-gain amplifier, providing virtual ground at nodes X and Y, and capacitors C_1 and C_2 track the input voltage. In the transition to the hold mode, first S_4 turns off, at which instant the input voltage is sampled on C_1 (and C_2), and subsequently S_1-S_3 turn off and S_5 turns on. Now, the circuit is configured as shown in Figure 3.10(b), where C_2, B_2, B_1, and C_1 form a unity-gain feedback loop around G_m, thus providing an output voltage approximately equal to the voltage stored on C_1.

It is instructive to study the sources of charge injection in this architecture. Since S_4 turns off with virtual ground at X and Y, it introduces a pedestal error independent of the input signal. The resulting offset is partially balanced by the pedestal due to the charge injected by S_2 onto C_3. Switch S_3, on the other hand, experiences swings equal to the input and injects an input-dependent charge onto nodes X_1 and Y_1 when it turns off. Since B_1 provides a low impedance at X_1, the charge injected onto this node gives no error. However, the charge injected onto Y_1 changes the voltage stored on C_2. Nonetheless, the negative feedback loop suppresses the effect of this change *at the output*. This occurs because the pedestal across C_2 propagates through B_2, B_1, and C_1 and appears at the inverting input of G_m, thereby causing the output of G_m to change in *the opposite* direction. As a result, the change in V_{out} is equal to the pedestal voltage across C_2 divided by the *voltage gain* of G_m, i.e., $G_{m2}R_{out}$, where R_{out} is the output resistance of G_m.

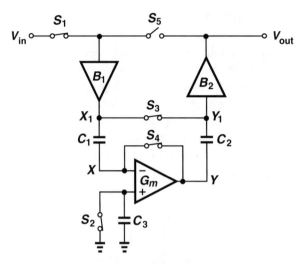

Fig. 3.9   Recycling architecture.

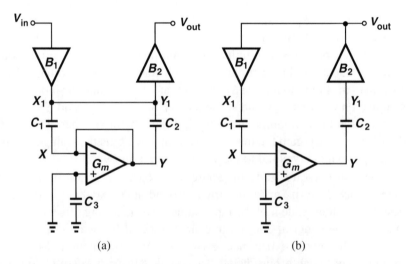

(a)   (b)

Fig. 3.10   Equivalent circuits of recycling architecture. (a) Sampling mode;
(b) hold mode.

In summary, the recycling architecture achieves small pedestal error, high linearity, and relatively high speed.

3.6 SWITCHED-CAPACITOR ARCHITECTURE

Evolved for use in A/D converters, the switched-capacitor architecture employs MOS switches extensively. Shown in Figure 3.11(a) in single-ended form, this architecture consists of input sampling capacitor C_H, transconductance amplifier G_m, and switches S_1-S_3 [12]. The circuit operates as follows. In the acquisition mode, S_1 and S_2 are on, S_3 is off, and G_m functions as a unity-gain amplifier, creating virtual ground at node X [Figure 3.11(b)]. Thus, the voltage across C_H tracks the input voltage. In the transition to the hold mode, first S_2 turns off, thereby sampling the instantaneous input voltage on C_H, and subsequently S_1 turns off and S_3 turns on. This results in the hold-mode configuration depicted in Figure 3.11(c), where C_H and G_m sustain an output voltage equal to the sampled input.

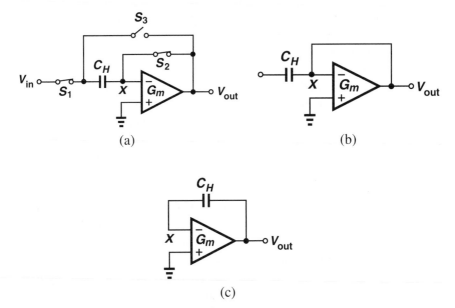

Fig. 3.11   Switched-capacitor SHA. (a) Basic circuit; (b) equivalent circuit in the acquisition mode; (c) equivalent circuit in the hold mode.

Since in this topology S_2 turns off first, the (input-dependent) charge injected by S_1 onto C_H does not appear in the held output voltage. Moreover, as S_2 is connected to virtual ground, its channel charge does not depend on the input signal. In a differential circuit, this charge would simply cause a common-mode offset.

The simplicity of this architecture has made it quite popular in applications such as pipelined A/D converters, where a large number of SHAs are required [12]. Since the dominant pole of the circuit is usually at the output, an increase in the load capacitance does not degrade the phase margin but may overcompensate the amplifier. Furthermore, a single-stage transconductance amplifier can be used to achieve linearities up to 13 bits [13].

In the architecture of Figure 3.11(a), the input switch S_1 experiences large voltage swings, introducing an input-dependent delay in the acquisition mode and hence harmonic distortion in the sampled signal. Furthermore, as the linearity and precision of the SHA strongly depend on the open-loop gain (the product of transconductance and output resistance) of G_m, the performance degrades if the circuit drives resistive loads, thereby limiting the use of the architecture to on-chip applications.

3.7 CURRENT-MODE ARCHITECTURE

Current-mode signal processing has been proposed as an alternative to the more conventional voltage-mode technique. Current-mode data conversion systems require current-input, current-output sampling circuits. However, even in this case the signal is stored as a voltage rather than a current because capacitors are far easier to fabricate than are inductors. Thus, in these architectures, first the input current must be converted to voltage so that it can be stored, and then the stored voltage must subsequently be converted to an output current.

The closed-loop architecture described in Section 3.2 can be easily modified to operate in the current mode, as shown in Figure 3.12. A variant of this architecture has been used in a current-mode A/D converter [14]. In this circuit, the inverting input of G_{m1} is grounded, the input current is summed with the output current of G_{m1} at node X, the signal is stored on C_H, and the output current is produced by G_{m2}. The closed-loop architecture is a natural choice for current-mode signals because, in the acquisition mode, it provides a virtual-ground summing node (X) as well as an internal current-to-voltage converter (consisting of A_0 and G_{m1}). Note that G_{m2} is outside the global feedback loop and its distortion is directly added to the stored signal.

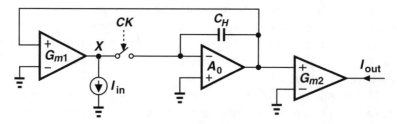

Fig. 3.12   Current-mode SHA derived from conventional closed-loop architecture.

The architecture of Figure 3.12 is topologically identical with the conventional closed-loop architecture and hence exhibits the same time response for both current-mode inputs and voltage-mode inputs. Consequently, the current-mode architecture faces the same stability issues as the voltage-mode architecture.

REFERENCES

[1] K. R. Stafford, et al., "A Complete Monolithic Sample/Hold Amplifier," *IEEE J. Solid-State Circuits,* vol. SC-9, pp. 381-387, Dec. 1974.

[2] M. J. Chambers and L. F. Linder, "A Precision Monolithic Sample-and-Hold for Video Analog-to-Digital Converters," *ISSCC Dig. Tech. Pap.,* pp. 168-169, Feb. 1991.

[3] K. Poulton, J. S. Kang, and J. J. Corcoran, "A 2 Gs/s HBT Sample and Hold," *Tech. Dig. 1988 GaAs IC Symp.,* pp. 199-202.

[4] P. Vorenkamp and J. P. M. Verdaasdonk, "Fully Bipolar, 120-Msample/s 10-b Track-and-Hold Circuit," *IEEE J. Solid-State Circuits,* vol. SC-27, pp. 987-992, July 1992.

[5] R. Jewett, J. J. Corcoran, and G. Steinbach, "A 12 b 20MS/sec Ripple-Through ADC," *ISSCC Dig. Tech. Pap.,* pp. 34-35, Feb. 1992.

[6] M. Nayebi and B. A. Wooley, "A 10 Bit Video BiCMOS Track and Hold Amplifier," *ISSCC Dig. Tech. Pap.,* pp. 68-69, Feb. 1989.

[7] P. J. Lim and B. A. Wooley, "A High-Speed Sample-and-Hold Technique Using a Miller Hold Capacitance," *IEEE J. Solid-State Circuits,* vol. SC-26, pp. 643-651, April 1991.

[8] C. R. Ryan, "Applications of a Four-Quadrant Multiplier," *IEEE J. Solid-State Circuits,* vol. SC-5, pp. 45-48, Feb. 1970.

[9] P. Petschacher et al., "A 10-b 75-MSPS Subranging A/D Converter with Integrated Sample and Hold," *IEEE J. Solid-State Circuits,* vol. SC-25, pp. 1339-1346, Dec. 1990.

[10] F. Moraveji, "A High-Speed Current-Multiplexed Sample-and-Hold Amplifier with Low Hold Step," *IEEE J. Solid-State Circuits,* vol. SC-26, pp. 1800-1808, Dec. 1991.

[11] P. Real and D. Mercer, "A 14 b Linear, 250 nsec Sample-and-Hold Subsystem with Self-Calibration," *ISSCC Dig. Tech. Pap.,* pp. 164-165, Feb. 1991.

[12] S. H. Lewis and P. R. Gray, "A Pipelined 5-Msample/s 9-bit Analog-to-Digital Converter," *IEEE J. Solid-State Circuits*, SC-22, pp. 954-961, Dec. 1987.

[13] Y. M. Lin, B. Kim, and P. R. Gray, "A 13-b 2.5-MHz Self-Calibrated Pipelined A/D Converter in 3-μm CMOS," *IEEE J. Solid-State Circuits*, vol. SC-26, pp. 628-636, April 1991.

[14] D. Robertson, P. Real, and C. Mangelsdorf, "A Wideband 10-bit, 20 Msps Pipelined ADC Using Current-Mode Signals," *ISSCC Dig. Tech. Papers*, pp. 160-161, Feb. 1990.

4

Basic Principles of Digital-to-Analog Conversion

Digital-to-analog conversion is an essential function in data processing systems. As mentioned in Chapter 1, D/A converters (DACs) interface the digital output of signal processors with the analog world. Moreover, as explained in Chapter 6, multistep analog-to-digital converters employ interstage DACs to reconstruct analog estimates of the input signal. Each of these applications imposes certain speed, precision, and power dissipation requirements on the DAC, mandating a good understanding of various D/A conversion techniques and their trade-offs.

In this chapter, we study the basic concepts and operations related to D/A conversion. Following a definition of performance metrics, we describe D/A conversion in terms of voltage, current, and charge division or multiplication and illustrate the merits and limitations of each approach. Finally, we discuss the switching functions needed to generate an analog output corresponding to a digital input.

4.1 GENERAL CONSIDERATIONS

A digital-to-analog converter produces an analog output A that is proportional to the digital input D:

$$A = \alpha D, \qquad (4.1)$$

where α is a proportionality factor. Since D is a dimensionless quantity, α sets both the dimension and the full-scale range of A. For example, if α is a current quantity, I_{REF}, then the output can be expressed as

$$A = I_{REF} D. \qquad (4.2)$$

In some cases, it is more practical to normalize D with respect to its full-scale value, 2^m, where m is the resolution. For example, if α is a voltage quantity, V_{REF},

$$A = V_{REF} \frac{D}{2^m}. \qquad (4.3)$$

From (4.2) and (4.3), we can see that in a D/A converter, each code at the digital input generates a certain multiple or fraction of a reference at the analog output. In other words, D/A conversion can be viewed as a reference multiplication or division function, where the reference may be one of the three electrical quantities: voltage, current, or charge. The accuracy of this function determines the linearity of the DAC, while the speed at which each multiple or fraction of the reference can be selected and established at the output gives the conversion rate of the DAC. Figure 4.1 shows the input/output characteristic of an ideal 3-bit D/A converter. The analog levels generated at the output follow a straight line passed through the origin and the full-scale point.

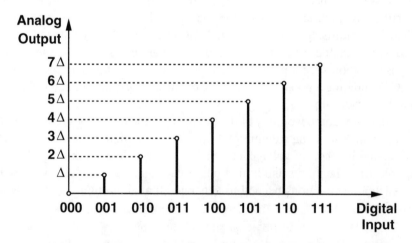

Fig. 4.1   Input/output characteristic of an ideal 3-bit D/A converter.

We should mention that in some applications such as "companding" DACs, the desired relationship between D and A is nonlinear [1], but in this book we discuss only "linear" or "uniform" DACs, i.e., those that ideally behave according to (4.2) or (4.3).

The digital input to a DAC can assume any predefined format but must eventually be of a form easily convertible to analog. Shown in Figure 4.2 are three formats often used in DACs: binary, thermometer, and 1-of-n codes. The latter two are shown in *column* form to make their visualization easier.

Decimal	0	1	2	3
Binary	00	01	10	11
Thermometer	0 0 0 0	0 0 0 1	0 0 1 1	0 1 1 1
1-of-n	0 0 0 0	0 0 0 1	0 0 1 0	0 1 0 0

Fig. 4.2   Binary, thermometer, and 1-of-n codes.

In the binary format, an m-bit number $D_{m-1} D_{m-2} \ldots D_0$ represents a decimal value of $D_{m-1} 2^{m-1} + D_{m-2} 2^{m-2} + \cdots + D_0 2^0$.

In the thermometer code, a number is represented by a column of j consecutive ONEs at the bottom and k consecutive ZEROs on top such that $j + k$ is a constant. For example, as shown in Figure 4.2, the decimal number 3 can be represented as three ONEs and one ZERO. This code can be viewed as a thermometer that is "filled" up to the topmost ONE in the column, and hence the name. We will also use the term "height" to refer to the number of ONEs in this code.

In the 1-of-n code, each number is represented as a single entry of ONE in a column of ZEROs, with the *position* of that entry showing the actual value. In Figure 4.2, for example, the decimal number 3 is depicted as a ONE in the third position from the bottom.

As seen from Figure 4.2, the thermometer and 1-of-n codes are much less compact than binary. Nonetheless, as discussed later, these codes are essential in D/A and A/D converter design.

4.2 PERFORMANCE METRICS

In this section, we define a number of terms usually used to characterize D/A converters. For a more complete set, the reader is referred to the literature

[2, 3] and manufacturers' data books. Figures 4.3 and 4.4 illustrate some of these metrics.

- Differential nonlinearity (DNL) is the maximum deviation in the output step size from the ideal value of one least significant bit (LSB).

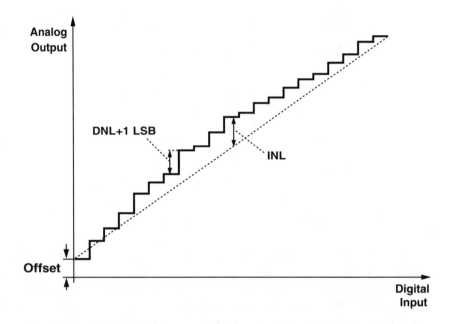

Fig. 4.3   Static parameters of D/A converters.

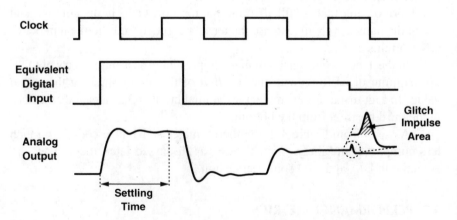

Fig. 4.4   Dynamic parameters of D/A converters.

- Integral nonlinearity (INL) is the maximum deviation of the input/output characteristic from a straight line passed through its end points. The difference between the ideal and actual characteristics will be called the INL profile.

- Offset is the vertical intercept of the straight line passed through the end points.

- Gain error is the deviation of the slope of the line passed through the end points from its ideal value (usually unity).

- Settling time is the time required for the output to experience full-scale transition and settle within a specified error band around its final value.

- Glitch impulse area is the maximum area under any extraneous glitch that appears at the output after the input code changes. This parameter is also called "glitch energy" in the literature even though it does not have an energy dimension.

- Latency is the total delay from the time the digital input changes to the time the analog output has settled within a specified error band around its final value. Latency may include multiples of the clock period if the digital logic in the DAC is pipelined.

- Signal-to-(noise + distortion) ratio (SNDR) is the ratio of the signal power to the total noise and harmonic distortion at the output when the input is a (digital) sinusoid.

Among these parameters, DNL and INL are usually determined by the accuracy of reference multiplication or division, settling time and delay are functions of output loading and switching speed, and glitch impulse depends on the D/A converter architecture and design.

Note that some of these parameters may be more important in some applications than in others. For example, many stand-alone DACs require low glitch area but may tolerate long latency. On the other hand, DACs utilized in A/D converters usually require a short latency but may have a relatively large glitch area.

4.3 REFERENCE MULTIPLICATION AND DIVISION

The linearity and SNDR of D/A converters strongly depend on the accuracy of the reference multiplication or division employed to generate the output levels. The three electrical quantities, voltage, current, and charge, can be multiplied or subdivided using resistor ladders, current-steering circuits, and switched-capacitor circuits, respectively. In this section, we describe each of these techniques and the errors that arise in typical implementations.

4.3.1 Voltage Division

A given reference voltage V_{REF} can be divided into N equal segments using a ladder composed of N identical resistors $R_1 = R_2 = \cdots = R_N$ (N is typically a power of 2) (Figure 4.5). An m-bit DAC requires a ladder with 2^m resistors, manifesting the exponential growth of the number of resistors as a function of resolution.

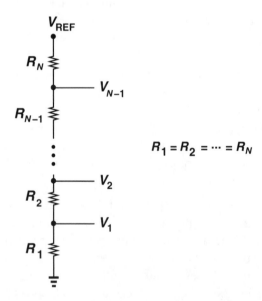

Fig. 4.5   Voltage division using a resistor ladder.

An important aspect of resistor ladders is the differential and integral nonlinearity they introduce when used in D/A converters. These errors result from mismatches in the resistors comprising the ladder.

In order to understand how resistor mismatch affects the DNL and INL of a resistor-ladder DAC, we first consider a simple case where the ladder exhibits a linear gradient, i.e., a linear variation in doping or width from one end to the other. This situation is shown in Figure 4.6. The voltage at the jth tap of this ladder is

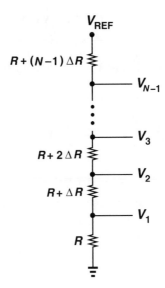

Fig. 4.6   Linear gradient in a resistor
ladder.

$$V_j = \frac{\sum\limits_{k=0}^{j-1}(R + k\,\Delta R)}{\sum\limits_{k=0}^{N-1}(R + k\,\Delta R)} V_{\text{REF}} \tag{4.4}$$

$$= \frac{jR + \dfrac{j(j-1)}{2}\,\Delta R}{NR + \dfrac{N(N-1)}{2}\,\Delta R} V_{\text{REF}}. \tag{4.5}$$

The INL profile is given by the difference between (4.5) and the ideal tap
voltage, $j V_{\text{REF}}/N$:

$$INL_j = \frac{j}{N} V_{\text{REF}} - \frac{jR + \dfrac{j(j-1)}{2}\,\Delta R}{NR + \dfrac{N(N-1)}{2}\,\Delta R} V_{\text{REF}}. \tag{4.6}$$

Simplifying (4.6) yields

$$INL_j = \frac{j(N-j)}{R + \dfrac{N-1}{2}\,\Delta R}\,\frac{\Delta R}{2N} V_{\text{REF}}. \tag{4.7}$$

Assuming $R \gg (N - 1)\Delta R/2$, we note that INL_j reaches a maximum of $N V_{REF}(\Delta R/8R)$ at $j = N/2$.

For the linear gradient depicted in Figure 4.6, the DNL is obtained by finding the deviation of $V_{j+1} - V_j$ from the ideal value of 1 LSB $(=V_{REF}/N)$:

$$\text{DNL}_j = V_{j+1} - V_j - \frac{V_{REF}}{N}, \tag{4.8}$$

which can be simplified to

$$\text{DNL}_j \approx (j - \frac{N - 1}{2}) \frac{\Delta R}{R} \frac{V_{REF}}{N} \tag{4.9}$$

if we assume $R \gg (N - 1) \Delta R/2$. For large N, the magnitude of this error reaches a maximum of approximately $V_{REF}(\Delta R/2R)$ at $j = 1$ and $j = N - 1$.

Note that if the maximum INL and DNL found above are to remain below 1 LSB, then the assumptions made in arriving at (4.7) and (4.9) are justified.

It is interesting to note that the nonlinearities described above are caused by perfectly *linear* resistors. This of course does not contradict any laws of linear systems because the switching operations in D/A converters make them inherently nonlinear systems.

While the case of linear gradients is simple and intuitive, in reality resistors also exhibit *random* mismatch. This type of mismatch originates primarily from uncertainties in geometry definition during processing, as well as random variations in contact resistance. Consider two resistors laid out with identical geometry and dimensions. In the ideal case, the value of each resistor can be expressed as

$$R = \frac{\rho L}{W t} + 2R_C, \tag{4.10}$$

where ρ is the resistivity, L, W, and t are the length, width, and thickness of the resistors, respectively, and R_C represents the additional resistance due to each contact. In reality, these resistors suffer from several mismatch components: resistivity mismatch, $\Delta\rho$, width, length, and thickness mismatch, ΔW, ΔL, and Δt, respectively, and contact resistance mismatch, ΔR_C. In a typical process, ΔW and ΔL result from limited edge definition capability in lithography and etching or deposition of the resistor material, Δt arises from gradients across the die, and ΔR_C is caused by random variations in the finite resistance at the interface of the resistor and the interconnect (usually metal).

Taking the total differential of (4.10), we can express the overall mismatch between the two resistors as

$$\Delta R = \frac{L\Delta\rho}{W t} + \frac{\rho\Delta L}{W t} - \frac{\rho L \Delta W}{W^2 t} - \frac{\rho L \Delta t}{W t^2} + 2\Delta R_C. \tag{4.11}$$

This value can be normalized to the mean value of the resistors, R, to yield the *relative* mismatch. Since in (4.10) $2R_C$ is usually a small fraction of R, the first four terms in (4.11) can be simplified by substituting $\rho L / W t$ for R:

$$\frac{\Delta R}{R} = \frac{\Delta \rho}{\rho} + \frac{\Delta L}{L} - \frac{\Delta W}{W} - \frac{\Delta t}{t} + 2\frac{\Delta R_C}{R}. \qquad (4.12)$$

Since contact resistance decreases as the resistor width increases, we can write $R_C = r/W$, where r is a proportionality factor. Thus, $\Delta R_C \approx \Delta r/W$, suggesting that (4.12) can be written as

$$\frac{\Delta R}{R} = \frac{\Delta \rho}{\rho} + \frac{\Delta L}{L} - \frac{\Delta W}{W} - \frac{\Delta t}{t} + 2\frac{t\Delta r}{\rho L}. \qquad (4.13)$$

In a typical process, ρ and t are given, leaving L, W, and R as the only variables under the designer's control. To minimize the overall mismatch, each of these parameters must be maximized. Of course, larger dimensions lead to higher parasitic capacitance between the resistor and the substrate, as well as larger chip area. Some properties of commonly available monolithic resistors are given in [4].

To analyze the nonlinearity resulting from random resistor mismatch, we must (1) assume a probability density function (PDF) for the value of each resistor, (2) calculate the resulting PDF for the tap voltages along the ladder, and (3) examine the mean and standard deviation of these voltages in terms of those of the resistors. Since random mismatch arises from a large number of random variables (corresponding to many uncorrelated events during fabrication), it is plausible to assume a Gaussian PDF for the resistor values. Using such an analysis, Kuboki et al. [5] have shown that the tap voltages of a resistor ladder follow a nearly Gaussian distribution, with a mean equal to $j V_{\text{REF}}/N$ and a standard deviation equal to

$$\Delta V_j = \sqrt{\frac{j}{N^2}\left(1 - \frac{j}{N}\right)} \, \frac{\Delta R}{R} V_{\text{REF}}, \qquad (4.14)$$

This error reaches a maximum of

$$\text{INL}_{\max} = \frac{1}{\sqrt{4N}} \, \frac{\Delta R}{R} V_{\text{REF}} \qquad (4.15)$$

at $j = N/2$. We should emphasize that this value is a standard deviation and hence a *likelihood* measure rather than a deterministic number. In other words, it implies that on the average, 68% of N-segment resistor ladders exhibit an error less than or equal to $(V_{\text{REF}}/\sqrt{4N})(\Delta R/R)$ at their midpoints. Figure 4.7 depicts an INL profile obtained by choosing each resistor of a 128-segment ladder from a Gaussian distribution with $R = 100 \ \Omega$ and $\Delta R = 5 \ \Omega$. We note that in this profile the maximum error *does not* occur at the midpoint and is actually *greater* than that predicted by (4.15).

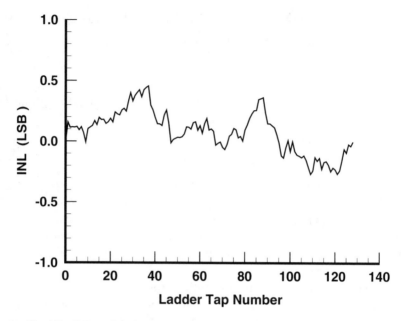

Fig. 4.7   INL profile of a ladder comprising segments chosen from a Gaussian
distribution.

Despite the counterexample of Figure 4.7, equation (4.15) does indicate
a trend: so long as $\Delta R / R$ remains constant, the maximum error decreases if
N increases. Thus, ladders with a large number of segments are more likely to
achieve a small (absolute) nonlinearity than are ladders with a small number
of segments. Intuitively, this is because random errors in the value of resistors
tend to average out when many segments are connected in series.

In addition to random mismatch, ladders made of diffused resistors exhibit a distinct nonlinear gradient. The thickness of the depletion layer under
these resistors is voltage-dependent and varies from one end of the ladder to
the other, thereby introducing a variation in the value of the resistor segments.
The resulting nonlinearity can be calculated by expressing the thickness of
the resistors as a square root function of the local voltage along the ladder.
Note that as the diffused resistor doping increases, this nonlinearity decreases,
whereas the depletion layer *capacitance*—which appears all along the ladder
to the substrate—increases. This trade-off makes diffused-resistor ladders
less attractive than poly-resistor ladders in applications where the ladder experiences transients and must recover quickly.

An important aspect of resistor ladder design is the Thevenin resistance
seen at each tap along the ladder. This resistance determines how fast capacitive loads charge or discharge to each tap's final voltage, contributing to

the settling time of the DAC. For an N-segment ladder, the Thevenin resistance reaches a maximum of $NR/4$ at the midpoint. This resistance grows exponentially with the number of bits but can be reduced at the cost of higher power dissipation.

Several variants of the resistor ladder shown in Figure 4.5 have been used for D/A conversion. We will describe these in Chapter 5.

4.3.2 Current Division

A reference current I_{REF} can be divided into N equal currents using N identical transistors connected as shown in Figure 4.8(a). (The same principle applies to both bipolar and MOS devices.) These currents can be combined so as to provide binary weighting, as depicted in Figure 4.8(b) for a 3-bit example. In this simple implementation, an m-bit DAC requires $2^m - 1$ transistors, resulting in a large number of devices for $m > 7$.

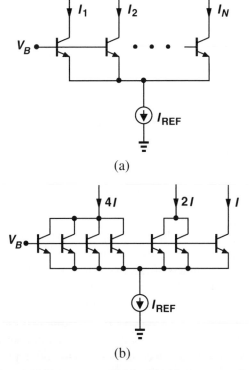

(a)

(b)

Fig. 4.8   (a) Uniform current division; (b) binary current division.

While conceptually simple, the implementation of Figure 4.8(a) has two drawbacks: the stack of current dividing transistors on top of I_{REF} limits the output voltage range, and I_{REF} must be N times each of the output currents; i.e., I_{REF} requires a very large device itself.

   These problems can be solved by current *replication* (rather than division) as shown in Figure 4.9(a), where N current mirrors generate N output currents *equal* to I_{REF}. In practice, to improve matching and linearity, degeneration resistors are placed in series with the emitters, yielding the circuit shown in Figure 4.9(b). This will be discussed later.

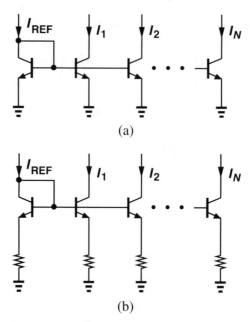

(a)

(b)

Fig. 4.9   (a) Simple current replication; (b) current replication incorporating emitter degeneration.

   As Figures 4.8(a) and (b) may suggest, current-steering arrays can be implemented in two different ways, using equal or binary-weighted current sources. Figure 4.10 depicts these two cases in a more general form. In the circuit of Figure 4.10(a), all current sources are equal and controlled by a thermometer code so that when the digital input increases by 1 LSB, one additional current source is switched to the output. In the circuit of Figure 4.10(b), the current sources are binary-weighted and controlled by a binary code so that each current source contributes to the output a current twice that of the next less significant bit. The configurations of Figures 4.10(a) and (b) are called "segmented" and "binary" arrays, respectively.

   An important feature of segmented arrays is their guaranteed monotonicity: since increments at the digital input simply cause an additive increment

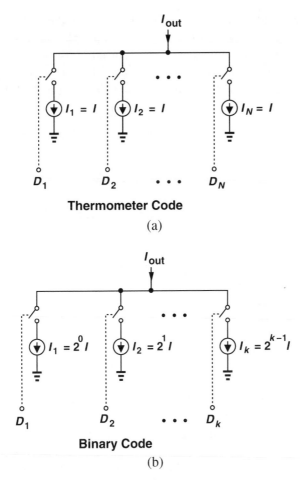

Fig. 4.10   (a) Segmented current-steering array; (b) binary current-steering array.

at the analog output, the transfer characteristic of such arrays is a monotonic function of the input, even if the maximum INL exceeds 1 LSB. In binary arrays, on the other hand, when the MSB current source [e.g., I_k in Figure 4.10(b)] turns on and all the current sources corresponding to lower bits turn off, the output may change by more than 1 LSB. As a result, in high-resolution applications such as strain gauge sensors, where the nonlinearity of the transducer itself is large and hence the converter linearity is not critical, segmented arrays are more attractive because their resolution (differential linearity) is relatively independent of their integral linearity.

The overall output current of a current-steering array can be converted to voltage using a resistor or a transimpedance amplifier, as shown in Figure

4.11. In Figure 4.11(a), a single resistor R_1 converts the current to voltage. This resistor can be a 50-Ω off-chip load if the DAC is to drive external circuits. In this case, the full-scale output current must be sufficiently large to produce reasonable voltage swings across R_1. Note that the output settling speed is limited by the total parasitic capacitance C_P at node X because this node experiences the entire output voltage swing.

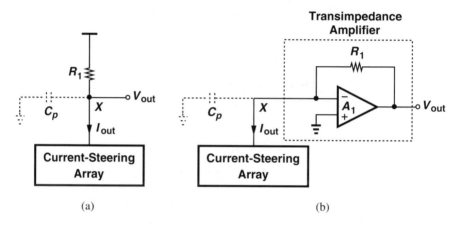

Fig. 4.11   Conversion of output current of an array to voltage using (a) a resistor and (b) a transimpedance amplifier.

In Figure 4.11(b), resistor R_1 is placed in a feedback loop around op amp A_1 to establish a virtual ground at node X. As a result, voltage variations at X are quite small and the output settling is determined by the op amp's speed. As discussed in Chapter 7, various trade-offs among speed, linearity, output voltage swing, and driving capability of op amps often restrict the dynamic range and settling speed of this topology. Consequently, the approach in Figure 4.11(a) is more common in high-speed applications.

Digital-to-analog converters that employ current-steering arrays suffer from three sources of nonlinearity: current source mismatch, finite output impedance of current sources (or the nonlinearity of the following transimpedance amplifier), and voltage dependence of the load resistor that converts the output current to voltage.

Current Source Mismatch.   The current sources in an array may exhibit mismatches due to gradients or random variations. The effect of gradients can be studied in the same fashion as for resistor ladders (Section 4.3.1). Hence, we consider only random mismatches here.

Consider two nominally identical current sources implemented as shown in Figure 4.12. The output currents I_1 and I_2 exhibit mismatch components due to mismatch between current gains and reverse saturation currents of Q_1 and Q_2 as well as mismatch between R_{E1} and R_{E2}. It can be shown [6] that the relative mismatch between I_1 and I_2 is

$$\frac{\Delta I}{I} \approx \frac{1}{1 + \dfrac{g_m R_E}{\alpha}} \frac{\Delta I_S}{I_S} + \frac{g_m R_E}{\alpha + g_m R_E}\left(\frac{\Delta \alpha}{\alpha} - \frac{\Delta R_E}{R_E}\right), \qquad (4.16)$$

where ΔI_S and $\Delta \alpha$ are the standard deviations of the reverse saturation current and common-base current gain of Q_1 and Q_2, respectively, ΔR_E is the standard deviation of R_{E1} and R_{E2}, and the same quantities without Δ represent the mean values. The transconductance of Q_1 and Q_2, $g_m \approx I/V_T$, where $V_T = kT/q$.

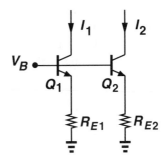

Fig. 4.12   Nominally identical bipolar current sources.

It is instructive to consider two extreme cases: (1) $R_{E1}, R_{E2} \ll 1/g_m$ and (2) $R_{E1}, R_{E2} \gg 1/g_m$. In the first case,

$$\frac{\Delta I}{I} \approx \frac{\Delta I_S}{I_S} \qquad (4.17)$$

i.e., the current mismatch is determined by transistor geometry matching. Since I_S is proportional to the emitter area of Q_1 and Q_2, its relative mismatch decreases as larger devices are used. For small transistors, this mismatch is on the order of 5 to 10%.

In the second case,

$$\frac{\Delta I}{I} \approx \frac{1}{g_m R_E} \frac{\Delta I_S}{I_S} + \left(\frac{\Delta \alpha}{\alpha} - \frac{\Delta R_E}{R_E}\right) \qquad (4.18)$$

$$\approx \frac{\Delta \alpha}{\alpha} - \frac{\Delta R_E}{R_E} \qquad (4.19)$$

i.e., current mismatch results from mismatch between current gains of Q_1 and Q_2 and between values of R_{E1} and R_{E2}.

Since the common-emitter current gain, β, is used more commonly than α, we note that $\alpha = \beta/(\beta + 1)$ and

$$\Delta\alpha = \frac{\Delta\beta}{(\beta + 1)^2}. \tag{4.20}$$

For $\beta \gg 1$, $\alpha \approx 1$ and

$$\frac{\Delta\alpha}{\alpha} \approx \frac{\Delta\beta}{\beta}\frac{1}{\beta}, \tag{4.21}$$

suggesting that relative mismatch in α is equal to relative mismatch in β divided by β. For a typical $\Delta\beta/\beta$ of 10% and $\beta = 100$, $\Delta\alpha/\alpha \approx 0.1\%$. As the second term in (4.19) is also on the order of 0.1%, we see that the output current mismatch in the second case is only a few tenths of percent, i.e., more than an order of magnitude lower than that in the first case.

It is important to note that in small-geometry transistors the emitter ohmic resistance (including contact and intrinsic components) is quite large and varies considerably from one device to another. This variation introduces significant mismatch between the output currents in the first case, whereas it adds to ΔR_E—and hence is suppressed by a factor of R_E—in the second case.

While increasing the size of the transistors reduces the error in (4.17), it also results in substantial collector-substrate and collector-base capacitance, thereby slowing down the switching.

For the above reasons, the second case is usually preferred. Another advantage of using emitter degeneration is higher output impedance, which reduces the integral nonlinearity as discussed later.

In CMOS technology, current sources are implemented using MOS transistors [Figure 4.13(a)]. Assume M_1 and M_2 are nominally identical and have square-law I-V characteristics:

$$I_D = \frac{1}{2}\mu C_{\text{ox}}\frac{W}{L}(V_{\text{GS}} - V_{\text{TH}})^2, \tag{4.22}$$

where μ is the carrier mobility in the channel, C_{ox} is the gate oxide capacitance per unit area, W and L are the device effective width and length, respectively, V_{GS} is the gate-source voltage, and V_{TH} is the threshold voltage. We can find the relative output current mismatch by taking the total differential of (4.22) and dividing the result by I_D:

$$\frac{\Delta I_D}{I_D} = \frac{\Delta(\mu C_{\text{ox}})}{\mu C_{\text{ox}}} + \frac{\Delta W}{W} - \frac{\Delta L}{L} - \frac{2\Delta V_{\text{TH}}}{V_{\text{GS}} - V_{\text{TH}}}. \tag{4.23}$$

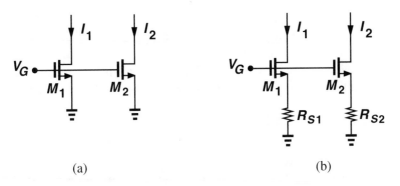

(a)   (b)

Fig. 4.13   Nominally identical MOS current sources.

Since μC_{ox} and V_{TH} are constant for a given process, W, L, and V_{GS} are the only parameters under the designer's control, and they can be increased to lower $\Delta I_D / I_D$. However, larger W leads to higher drain-substrate and gate-drain capacitance and larger area, larger L requires higher $V_{GS} - V_{TH}$ to attain a given I_D, and increasing $V_{GS} - V_{TH}$ limits the voltage swing at the drain of M_1 and M_2. As a consequence, some compromise is usually necessary to obtain a reasonable combination of accuracy, speed, and output voltage swing. Also, note that as W and L increase, both $\Delta(\mu C_{ox})$ and ΔV_{TH} tend to average out over the gate area of the transistors and hence become smaller [7].

For short-channel MOSFETs ($L < 2\,\mu m$), the I-V characteristics deviate from the square law because of velocity saturation, mobility degradation due to vertical field, and threshold voltage variation with drain-source voltage. As an extreme case, let us assume very short channels and heavy velocity saturation. Then,

$$I_D \approx W C_{ox}(V_{GS} - V_{TH})v_{sat},\qquad(4.24)$$

where v_{sat} is the saturation velocity of carriers [8]. Thus,

$$\frac{\Delta I_D}{I_D} = \frac{\Delta(C_{ox}v_{sat})}{C_{ox}v_{sat}} + \frac{\Delta W}{W} - \frac{\Delta V_{TH}}{V_{GS} - V_{TH}},\qquad(4.25)$$

indicating the same trends as (4.23) except that the mismatch is independent of L.

For MOS current sources, it is possible to use source degeneration resistors as shown in Figure 4.13(b) [9]. However, this improves the matching significantly only if the value of these resistors is comparable with the inverse of the transconductance of the MOSFETs. To understand why, note that in Figure 4.13(b) we can write

$$I_D R_S + \sqrt{\frac{2I_D}{\mu C_{ox} W/L}} + V_{TH} = V_G, \tag{4.26}$$

where all quantities are mean values. Taking the total differential of both sides and substituting $g_m = \sqrt{2\mu C_{ox} W I_D/L} = 2I_D/(V_{GS} - V_{TH})$, we have

$$\frac{\Delta I_D}{I_D} = \frac{1}{1 + g_m R_S}[\frac{\Delta(\mu C_{ox})}{\mu C_{ox}} + \frac{\Delta W}{W} - \frac{\Delta L}{L} - \frac{2\Delta V_{TH}}{V_{GS} - V_{TH}} - g_m \Delta R_S]. \tag{4.27}$$

The first four terms in the square brackets are the same as those in (4.23), but their effect is divided by $1 + g_m R_S$. The last term represents additional mismatch contributed by R_{S1} and R_{S2} themselves. This equation indicates that source degeneration is effective only if $g_m R_S$ is comparable with (and preferably much greater than) unity. Since for given V_G and I_D, the value of R_S cannot exceed $(V_G - V_{TH})/I_D$, a large $g_m R_S$ means a high g_m, which in turn demands wide transistors.

The effect of current source mismatch on integral linearity can be analyzed in the same manner as in the case of resistor ladders (Section 4.3.1). The important result is that as the number of current sources in the array increases, the relative nonlinearity decreases because random errors tend to average out.

Finite Output Impedance of Current Sources.   If the output current-summing node of a current-steering array experiences large voltage excursions [e.g., Figure 4.11(a)], then another type of integral nonlinearity arises from the finite output impedance of the current sources. Intuitively we note that, as the output varies between zero and full-scale, different impedances are switched to the output node, thereby introducing variations in the equivalent load resistance and hence nonlinearity in the output voltage. As an example, we calculate this nonlinearity for a segmented array.

Consider the small-signal equivalent circuit shown in Figure 4.14, where r_O represents the output impedance of each current source. For a thermometer code of height j, the output voltage is

$$V_{out} = -jI(R_1||\frac{r_O}{j}). \tag{4.28}$$

The dependence of the term in parentheses introduces integral nonlinearity. To obtain the INL profile, we pass a straight line through the end points of (4.28) (given by $j = 0$ and $j = N$) and find the difference between (4.28) and that line. Assuming $r_O \gg N R_1$, the reader can easily show that

$$\text{INL}_j \approx \frac{I R_1^2}{r_O} j(N - j), \tag{4.29}$$

which has a maximum of $I R_1^2 N^2 / 4 r_O$. Since the full-scale output voltage is approximately equal to $N I R_1$, the maximum relative nonlinearity is equal to $N R_1 / 4 r_O$. As this value must be very small, the assumption made in arriving at (4.29) is valid.

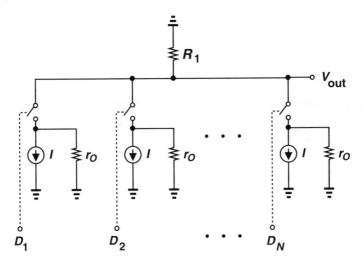

Fig. 4.14 Current-steering array including output impedance of each current source.

Load Resistor Nonlinearity.   Another type of nonlinearity in current-steering arrays stems from voltage dependence of the load resistor [R_1 in Figures 4.11(a) and (b)]. Polysilicon resistors exhibit a hyperbolic sine *I-V* characteristic that becomes more linear as the length of the resistor increases [10]. In diffused resistors, on the other hand, nonlinearity originates from the variation of the width of the underlying depletion region and is primarily a function of doping levels.

For linearities above 10 bits, these effects and their temperature dependence must be accurately measured and characterized for each process.

4.3.3 Charge Division

A reference charge, Q_{REF}, can be divided into N equal packets using N identical capacitors configured as in Figure 4.15. In this circuit, before S_1 turns on, C_1 has a charge equal to Q_{REF}, while $C_2, \ldots, C_N$ have zero charge. When S_1 turns on, Q_{REF} is distributed equally among $C_1, \ldots, C_N$, yielding a charge of Q_{REF}/N on each. Further subdivision can be accomplished by disconnecting one of the capacitors from the array and redistributing its charge among some other capacitors.

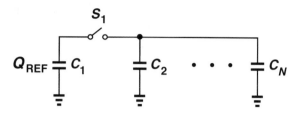

Fig. 4.15   Simple charge division.

While the circuit of Figure 4.15 can operate as a D/A converter if a separate array is employed for each bit of the digital input, the resulting complexity prohibits its use for resolutions above 6 bits. A modified version of this circuit is shown in Figure 4.16(a). Here, identical capacitors $C_1 = \cdots = C_N = C$ share the same top plate, and their bottom plates can be switched from ground to a reference voltage, V_{REF}, according to the input thermometer code. In other words, each capacitor can inject a charge equal to CV_{REF} onto the output node, thereby producing an output voltage proportional to the height of the thermometer code. The circuit operates as follows. First, S_P is on and the bottom plates of $C_1, \ldots, C_N$ are grounded, discharging the array to zero [Figure 4.16(b)]. Next, S_P turns off, and a thermometer code with height j is applied at $D_1, \ldots, D_N$, thus connecting the bottom plate of $C_1, \ldots, C_j$ to V_{REF} and generating an output equal to jV_{REF}/N [Figure 4.16(c)].

The circuit of Figure 4.16 is, in the strict sense, a voltage divider rather than a charge divider. In fact, the expression relating its output voltage to V_{REF} and the value of the capacitors is quite similar to that of resistor ladders. Nonetheless, in considering nonlinearity and loading effects it is helpful to remember that the circuit's operation is based on charge injection and redistribution.

As with current-steering DACs, capacitor arrays can be configured in either segmented or binary form. The example considered in Figure 4.16 is a segmented array, with the important property of having a monotonic input/output characteristic. Shown in Figure 4.17 is a binary version, where the digital input assumes a binary format and unit capacitors are grouped so as to provide binary weighting. The operation is similar to that of Figure 4.16.

Nonlinearity of capacitor DACs arises from three sources: capacitor mismatch, capacitor nonlinearity, and the nonlinearity of the junction capacitance of any switches connected to the output node.

Capacitor Mismatch.   Mismatch in an array of capacitors arises from both gradients and random variations. The effect of gradients can be formu-

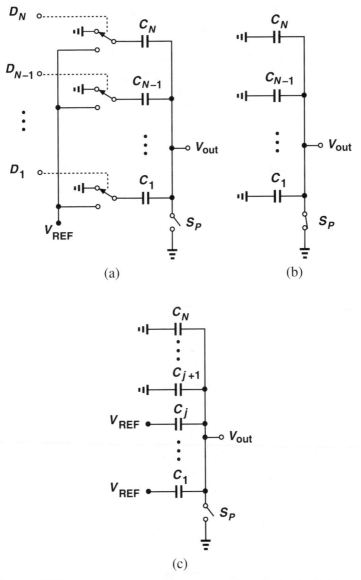

Fig. 4.16   (a) Modified charge division configuration; (b) circuit of (a) in discharge mode; (c) circuit of (a) in evaluate mode.

lated in the same manner as in the case of resistor ladders (Section 4.3.1). Thus, we discuss only random variations here.

Consider two nominally identical monolithic capacitors. Due to dimension and oxide thickness mismatch, these capacitors exhibit a relative

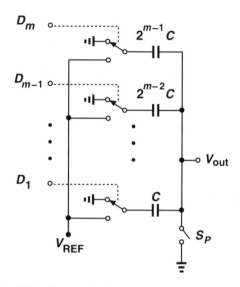

Fig. 4.17   Charge division with binary digital inputs.

mismatch of

$$\frac{\Delta C}{C} = \frac{\Delta W}{W} + \frac{\Delta L}{L} - \frac{\Delta t_{ox}}{t_{ox}}, \tag{4.30}$$

where $\Delta W/W$, $\Delta L/L$, and $\Delta t_{ox}/t_{ox}$ are relative mismatches in the width, length, and oxide thickness, respectively. This expression indicates that increasing W, L, or t_{ox} improves the matching. In practice, t_{ox} is a constant of the process, leaving W and L as the only variables. However, increasing W and L does not reduce $\Delta C/C$ indefinitely because t_{ox} gradients become more significant at large dimensions and raise the third term in (4.30). As a consequence, $\Delta C/C$ reaches a minimum at certain optimum dimensions [12]. Although process-dependent, these dimensions are usually in the range of 20 to 30 μm. To improve the matching between large capacitors, common-centroid geometries can be used [11].

The effect of random capacitor mismatch on the overall linearity can be analyzed in a fashion similar to that for resistor ladders in Section 4.3.1. We note that the integral linearity improves as the number of capacitors in the array increases because random errors tend to average out.

Capacitor Voltage Dependence.   Another source of nonlinearity in capacitor DACs is capacitor voltage dependence [11]. Figure 4.18 shows the structure of a typical monolithic capacitor consisting of two (doped) polysilicon plates separated by a dielectric (usually silicon dioxide). The voltage dependence originates from the variation of both the dielectric constant, ϵ, and the thickness of the depletion region within each plate [11, 13]. Since the actual expressions for this dependence are often quite complex, it is common practice to model the nonlinearity with a polynomial (i.e., Taylor expansion) such as

$$C = C_0 + C_0\alpha_1 V + C_0\alpha_2 V^2 + \cdots, \qquad (4.31)$$

where α_j is the jth-order voltage coefficient of the capacitor. It is important to note that the representation in (4.31) *does not* mean that the total charge on a nonlinear capacitor is equal to the product of its voltage and the value of the capacitor at that voltage. Rather, (4.31) simply shows that if the capacitor has a voltage V, then an *increment* of voltage, dV, requires an *increment* of charge, dq, given by

$$dq = C_0(1 + \alpha_1 V + \alpha_2 V^2 + \cdots)\,dV. \qquad (4.32)$$

Consequently, the total charge needed to change the voltage of a capacitor from V_1 to V_2 is

$$\Delta Q = C_0 \int_{V_1}^{V_2} (1 + \alpha_1 V + \alpha_2 V^2 + \cdots)\,dV. \qquad (4.33)$$

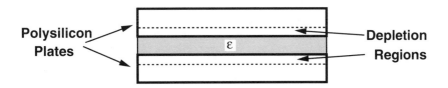

Fig. 4.18 Typical monolithic capacitor structure.

For most applications, the first three terms in (4.31) represent the capacitor nonlinearity with reasonable accuracy. In single-ended circuits, α_1 is usually the dominant factor, whereas in differential configurations the effect of α_2 is significant because the even-order nonlinearity components resulting from α_1 are suppressed.

As an example of computing INL due to capacitor nonlinearity, we analyze the effect of α_1 on a segmented capacitor array. For an input thermometer code of height j, the equivalent circuit of Figure 4.19 can be constructed. Here, the charge on C_1 must equal that on C_2, and as the output goes from zero to its final value, the voltage across C_1 changes from zero to $V_{\text{REF}} - V_{\text{out}}$ and that across C_2 from zero to V_{out}. It follows from (4.33) that

$$\int_0^{V_{\text{out}}} (N - j)C_0(1 + \alpha_1 V)\,dV = \int_0^{V_{\text{REF}} - V_{\text{out}}} jC_0(1 + \alpha_1 V)\,dV. \quad (4.34)$$

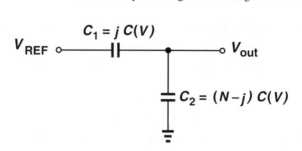

Fig. 4.19   Equivalent circuit of a segmented capacitor array with an input thermometer code of height j.

Carrying out the integration and solving for $j V_{\mathrm{REF}}/N$ (the ideal input/output characteristic), we have

$$\frac{j}{N} V_{\mathrm{REF}} = \frac{\alpha_1 V_{\mathrm{out}} + 2}{2 + \alpha_1 V_{\mathrm{REF}} - 2\alpha_1 V_{\mathrm{out}} + 2\alpha_1 V_{\mathrm{out}}^2/V_{\mathrm{REF}}} V_{\mathrm{out}}. \qquad (4.35)$$

For typical values of $\alpha_1 < 100$ ppm/V and V_{REF} and $V_{\mathrm{out}} \leq 5$ V, the last three terms in the denominator of (4.35) are much less than 2 and hence

$$\frac{j}{N} V_{\mathrm{REF}} \approx V_{\mathrm{out}} - \frac{\alpha_1 V_{\mathrm{out}}}{2}\left(V_{\mathrm{REF}} - 3V_{\mathrm{out}} + \frac{2V_{\mathrm{out}}^2}{V_{\mathrm{REF}}}\right). \qquad (4.36)$$

This equation can be rewritten as

$$V_{\mathrm{out}} \approx \frac{j}{N} V_{\mathrm{REF}} + \frac{\alpha_1 V_{\mathrm{out}}}{2}\left(V_{\mathrm{REF}} - 3V_{\mathrm{out}} + \frac{2V_{\mathrm{out}}^2}{V_{\mathrm{REF}}}\right). \qquad (4.37)$$

The second term vanishes at $V_{\mathrm{out}} = 0$ and $V_{\mathrm{out}} = V_{\mathrm{REF}}$, and it represents the nonlinearity in the transfer characteristic. This term reaches a maximum of

$$\mathrm{INL}_{\mathrm{max}} = \frac{\sqrt{3}}{36}\alpha_1 V_{\mathrm{REF}}^2 \qquad (4.38)$$

at $V_{\mathrm{out}} = (3 \pm \sqrt{3})V_{\mathrm{REF}}/6$.

   The above analysis can be repeated for binary arrays in a similar fashion [14].

Switch Junction Capacitance Nonlinearity.   The third source of INL in capacitor DACs is the nonlinearity of the junction capacitance of the switch(es) connected to the output node (e.g., S_P in Figures 4.16 and 4.17). The variation of source or drain junction capacitance of MOSFETs can be expressed as

$$C_{\mathrm{junc}} = \frac{C_0}{(1 + V_j/\Phi)^{m_j}}, \qquad (4.39)$$

where C_0 is the zero-bias capacitance, V_j is the voltage across the junction, Φ is the built-in potential, and m_j is typically between 0.3 and 0.5. For analysis purposes, we can approximate (4.39) with an empirical polynomial such as

$$C_{junc} \approx C_0[0.1m_j(\frac{V_j}{\Phi})^2 - 0.63m_j(\frac{V_j}{\Phi}) + 1]. \qquad (4.40)$$

This relation can be used to estimate the nonlinearity for a given voltage swing at the output. The analysis is similar to that performed above.

In capacitor DACs, the top plate parasitic capacitance of the array introduces gain error. As shown in Figure 4.20(a), this capacitance results from electric field lines that emanate from the top plate and terminate on the substrate rather than the bottom plate. With this component, a segmented

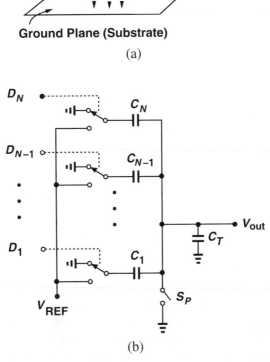

Ground Plane (Substrate)

(a)

(b)

Fig. 4.20   (a) Top plate parasitic of a capacitor; (b) segmented capacitor array including top plate parasitic.

capacitor DAC, for example, can be represented as in Figure 4.20(b), with C_T modeling the total top plate parasitic. Here, the output voltage corresponding to a thermometer code of height j is

$$V_0 = \frac{jC}{NC + C_T} V_{REF}. \tag{4.41}$$

Since typically $NC \gg C_T$, (4.41) is simplified as

$$V_0 = \frac{jC}{NC[1 + C_T/(NC)]} V_{REF} \tag{4.42}$$

$$\approx \frac{j}{N}(1 - \frac{C_T}{NC}) V_{REF} \tag{4.43}$$

indicating that the gain deviates from unity by $C_T/(NC)$.

While in many stand-alone D/A converters gain error is not a serious drawback (and can often be nulled by external means), in DACs used in multistep A/D converters it becomes crucial because it leads to differential nonlinearity. This issue is discussed in Chapter 6.

4.4 SWITCHING AND LOGICAL FUNCTIONS IN DACS

In order to produce at the output of a DAC an analog quantity proportional to the digital input, some switching and logical functions are required. In this section, we describe a number of these functions often employed in the design of D/A converters. In Chapter 5, we see how these functions are chosen in different architectures.

4.4.1 Switching Functions in Resistor-Ladder DACs

In resistor-ladder DACs with binary inputs, the switching function can be implemented as an analog multiplexer [15]. As depicted in Figure 4.21(a), the binary input selects one of the ladder taps, providing a resistive path from that tap to the output. Thus, if no current is drawn from the output, V_{out} is equal to the voltage of the selected tap. Note that for an m-bit DAC, m MOS switches appear in series between each ladder tap and the output node. The total channel resistance of these switches often limits the settling speed at the output and raises concern for the thermal noise added to V_{out}. Furthermore, the output node must be buffered by a high input impedance amplifier because resistive loads directly connected to V_{out} cause attenuation and nonlinearity (as the on-resistance of switches varies with the output voltage). In addition,

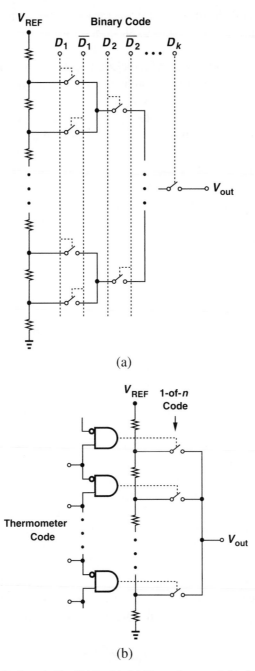

(a)

(b)

Fig. 4.21   Resistor-ladder DAC with (a) binary input and (b) thermometer code input.

if V_{REF} is comparable with the gate overdrive of the switches, the topmost switches in Figure 4.21(a) exhibit substantial on-resistance unless they use complementary devices.

For resistor-ladder DACs with thermometer code inputs—such as those utilized in A/D converters—the switching can be realized as shown in Figure 4.21(b). Here, NAND gates are used to compare every two adjacent bits of the thermometer code, thereby detecting the topmost ONE and generating a 1-of-n code. This code then turns on the corresponding switch, making V_{out} equal to a ladder tap voltage. In contrast with the multiplexing configuration of Figure 4.21(a), this DAC uses only one switch between each ladder tap and the output, thus achieving a lower overall channel resistance and more modularity in layout. However, for an m-bit DAC with thermometer code input, 2^m switches are connected to the output node, introducing considerable junction and overlap capacitance at that node. For the same DAC with binary inputs, on the other hand, the capacitance of only $2m$ switches appears in the path from ladder taps to the output.

Other ladder switching topologies are described in the context of DAC architectures in Chapter 5.

4.4.2 Switching Functions in Current-Steering DACs

Current-steering DACs configured as segmented or binary arrays often use a unit current-steering cell. Figure 4.22 illustrates different current-switching cells in bipolar and CMOS technologies. Note that these circuits can naturally provide differential outputs.

A comparison of the speed and precision of these cells can reveal their merits and drawbacks. The exponential I-V characteristic of bipolar transistors allows complete switching of current with only a few hundred millivolts of differential swing at D and $\overline{D}$ in Figure 4.22(a). On the other hand, the MOS pair in Figure 4.22(b) typically requires differential swings of more than 1 V at D and $\overline{D}$ to ensure that one side turns off completely. This difference in swings indicates two advantages for bipolar switches over MOS devices. First, the input voltage transition is faster. Second, the voltage swing at node X in Figure 4.22(a) is smaller and hence the feedthrough to the bias line V_b is less.

An important aspect of the above current-steering cells is their output impedance because it contributes to the overall integral nonlinearity (Section 4.3.2). For the bipolar cell, this impedance is [6]

$$R_{out} = \frac{1 + g_{m3}R_E}{1 + \dfrac{g_{m3}R_E}{\beta}} r_{O3} g_{m12} r_{O12}, \qquad (4.44)$$

where g_{m3} is the transconductance of Q_3, and g_{m12} and r_{O12} are the transcon-

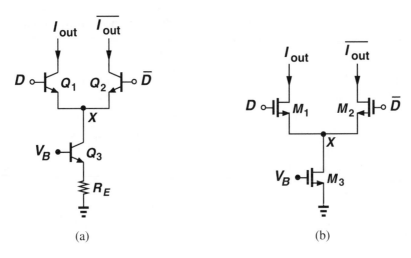

Fig. 4.22   Current-steering cells in (a) bipolar and (b) MOS technologies.

ductance and output impedance of Q_1 or Q_2 (whichever is on), respectively. This equation can be rewritten as

$$R_{\text{out}} = \frac{V_{RE} + V_T}{V_{RE} + \beta V_T} \beta r_{O3} g_{m12} r_{O12}, \qquad (4.45)$$

where V_{RE} denotes the voltage drop across R_E. In a typical case, $V_T \ll V_{RE} \ll \beta V_T$, yielding:

$$R_{\text{out}} \approx \frac{V_{RE}}{V_T} r_{O3} g_{m12} r_{O12}. \qquad (4.46)$$

Thus, emitter degeneration increases the output impedance by a factor of V_{RE}/V_T, making it desirable to maximize this ratio in the presence of voltage headroom constraints. Note that as explained in Section 4.3.2, increasing this ratio also lowers mismatch between cells by suppressing the effect of emitter area mismatch.

For the MOS current-steering cell, the output impedance depends on whether M_1 and M_2 are in the saturation or the linear region when they are on. If these devices are in saturation, then

$$R_{\text{out}} \approx r_{O3} g_{m12} r_{O12}. \qquad (4.47)$$

For a given current, this value can be increased by increasing the length of M_1-M_3 or the width of M_1 and M_2. To maximize speed, M_1 and M_2 usually employ minimum length and M_3 is sized to provide adequate output impedance.

If M_1 and M_2 in Figure 4.22(b) enter the deep linear region, then

$$R_{out} = r_{O3} + R_{on12},\tag{4.48}$$

where R_{on12} is the on-resistance of M_1 and M_2. This value is much less than that in (4.47) and often inadequate for current-steering DACs.

Another consideration in high-resolution DACs is that the settling time at the output node usually dominates the conversion speed. Consequently, the current switching delay is only a small fraction of the total conversion time, making the difference between switching speeds of bipolar and MOS pairs negligible.

It is important to note that the output voltage swing of current steering DACs is quite limited, whereas it can extend to both supply rails in resistor-ladder and capacitor DACs, making the latter two more suitable when a wide dynamic range is necessary.

Other variants of current-steering cells are discussed in [16, 17].

4.4.3 Switching Functions in Capacitor DACs

Figure 4.23 illustrates the implementation of switching function in a typical capacitor DAC. Depending on the digital input format (binary or thermometer code), the devices in this circuit have uniform or binary weighting. The digital input controls the MOS devices that switch the bottom plate of the capacitors between ground and V_{REF}. The top plate is discharged to ground by S_P during reset.

It is instructive to calculate the settling time of this DAC as a function of the sizes of the capacitors, the switches, and the load capacitance. As an example, we consider a segmented array, shown in simplified form in Figure 4.24 for a thermometer input of height j. Here, R_k represents the on-resistance of the kth switch and C_L denotes the load capacitance. Assuming $R_1 = \cdots = R_N = R$, $C_1 = \cdots = C_N = C$, and $NC \gg C_L$, the reader can easily show that the output settling behavior is described by a single time constant equal to

$$\tau_{out} = \frac{RC_L}{N}.\tag{4.49}$$

Note that τ_{out} is *independent* of the value of the capacitors in the array. This is because all the devices in the array were assumed identical, thereby resulting in cancellation of higher-order terms of the complex frequency s. In reality, the capacitors and switches do not match perfectly, yielding additional components in the settling waveform. Equation (4.49) nonetheless shows that the

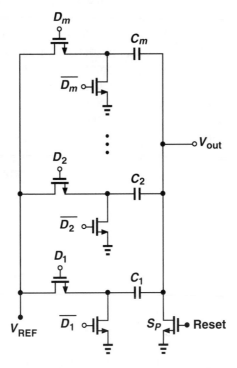

Fig. 4.23   Switching in a capacitor DAC.

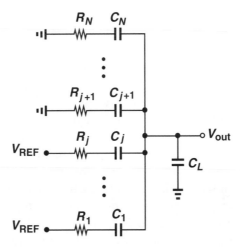

Fig. 4.24   Simplified small-signal model of a segmented capacitor array.

time constant can be quite small because the on-resistances of all the switches appear *in parallel* when the circuit drives C_L.

Another issue in this DAC relates to the size of the reset switch S_P. In order to suppress both the nonlinearity of source or drain junction capacitance and the offset caused by charge injection of S_P, its size must be minimized. However, this would result in a very slow reset if S_P were to discharge the entire array to ground. This problem can be avoided if the bottom plates of capacitors are grounded during reset so that the array capacitors remove the charge that they previously injected onto the output node in the conversion mode [18].

In the capacitor DACs used in successive approximation A/D converters, the voltage of the top plate of the array always returns to zero at the end of conversion (Chapter 6). Thus, in such applications the nonlinearity of the junction capacitance of the top plate switch is negligible.

The switching operation in capacitor DACs often draws very large transient currents from V_{REF}. For example, if 64 capacitors in a 128-unit array switch from ground to $V_{REF} = 2$ V and if each capacitor has a value of 0.5 pF and each switch an on-resistance of 1 kΩ, then the initial current drawn from V_{REF} exceeds 50 mA and it reaches this peak in only a few nanoseconds. If V_{REF} is supplied externally, the lead and trace inductance of the package give rise to substantial transient voltage drops as a result of such a large current spike, thus increasing the output settling time dramatically. If V_{REF} is generated on the chip, the reference generator must exhibit low output impedance *even for high-speed transients*, a requirement often difficult to meet with reasonable power dissipation.

4.4.4 Binary-to-Thermometer Code Conversion

Various features of the thermometer code have made it attractive for use in D/A converters. Since in stand-alone DACs the digital input is typically binary, a binary-thermometer code conversion is usually necessary.

Shown in Figure 4.25 are the binary and thermometer codes corresponding to decimal numbers 0 to 7 along with logical expressions that perform this conversion. These expressions can be easily implemented using one level of gates.

While the relationships in Figure 4.25 can be realized as random logic, it is advantageous to create a modular form applicable to higher word lengths. In most stand-alone DACs, the logic is implemented in several levels, often with pipelining in between. We discuss these techniques in Chapter 5.

Binary   **Thermometer**

A B C	T_1 T_2 T_3 T_4 T_5 T_6 T_7
0 0 0	0 0 0 0 0 0 0
0 0 1	1 0 0 0 0 0 0
0 1 0	1 1 0 0 0 0 0
0 1 1	1 1 1 0 0 0 0
1 0 0	1 1 1 1 0 0 0
1 0 1	1 1 1 1 1 0 0
1 1 0	1 1 1 1 1 1 0
1 1 1	1 1 1 1 1 1 1

$$T_1 = A + B + C \qquad T_5 = A\,(B + C)$$
$$T_2 = A + B \qquad T_6 = AB$$
$$T_3 = A + BC \qquad T_7 = ABC$$
$$T_4 = A$$

Fig. 4.25   Binary-to-thermometer code conversion.

REFERENCES

[1] Y. P. Tsividis et al., "A Segmented μ-255 Law PCM Voice Encoder Utilizing NMOS Technology," *IEEE J. Solid-State Circuits*, vol. SC-11, pp. 740-747, Dec. 1979.

[2] S. K. Tewksbury et al., "Terminology Related to the Performance of S/H, A/D, and D/A Circuits," *IEEE Trans. Circuits Syst.*, vol. CAS-25, pp. 419-426, July 1978.

[3] IEEE Standard 746-1984, "Performance Measurements of A/D and D/A Converters for PCM Television Video Circuits," IEEE, New York, 1984.

[4] D. J. Allstot and W. C. Black, "Technological Design Considerations for Monolithic MOS Switched Capacitor Filtering Systems," *Proc. IEEE*, vol. 71, pp. 967-986, Aug. 1983.

[5] S. Kuboki et al., "Nonlinearity Analysis of Resistor String A/D Converters," *IEEE Trans. Circuits Syst.*, vol. CAS-29, pp. 383-389, June 1982.

[6] P. R. Gray and R. G. Meyer, *Analysis and Design of Analog Integrated Circuits*, Third Edition, John Wiley and Sons, New York, 1992.

[7] M. J. M. Pelgrom, A. C. J. Duinmaijer, and A. P. G. Welbers, "Matching Properties of MOS Transistors," *IEEE J. Solid-State Circuits*, vol. SC-24, pp. 1433-1439, Oct. 1989.

[8] S. M. Sze, *Physics of Semiconductor Devices*, Second Edition, John Wiley and Sons, New York, 1981.

[9] K. R. Lakshmikumar et al., "A High-Speed 8-Bit Current Steering CMOS DAC," *Proc. CICC*, pp. 156-159, May 1985.

[10] N. C. C. Lu et al., "Modeling and Optimization of Monolithic Polycrystalline Silicon Resistors," *IEEE Trans. Electron. Devices*, vol. ED-28, pp. 818-830, July 1981.

[11] J. L. McCreary, "Matching Properties, and Voltage and Temperature Dependence of MOS Capacitors," *IEEE J. Solid-State Circuits*, vol. SC-16, pp. 608-616, Dec. 1981.

[12] J. B. Shyu, G. C. Temes, and F. Krummenacher, "Random Error Effects in Matched MOS Capacitors and Current Sources," *IEEE J. Solid-State Circuits*, vol. SC-19, pp. 948-955, Dec. 1984.

[13] C. Kaya et al., "Polycide/Metal Capacitors for High Precision A/D Converters," *IEDM Dig. Tech. Pap.*, pp. 782-785, Dec. 1988.

[14] H. S. Lee and D. A. Hodges, "Accuracy Considerations in Self-Calibrating A/D Converters,"*IEEE Trans. Circuits Syst.*, vol. CAS-32, pp. 590-597, June 1985.

[15] A. R. Hamade, "A Single Chip All-MOS 8-Bit A/D Converter," *IEEE J. Solid-State Circuits*, vol. SC-13, pp. 785-791, Dec. 1979.

[16] T. Miki et al., "A 10 Bit 50 MS/s CMOS D/A Converter with 2.7 V Power Supply," *Proc. VLSI Circuits Symp.*, pp. 92-93, June 1992.

[17] N. Kumazawa et al., "An 8 bit 150 MHz D/A Converter with 2.2 V Wide Range Output," *Proc. VLSI Circuits Symp.*, pp. 55-56, June 1990.

[18] B. Razavi and B. A. Wooley, "A 12-b 5-Msample/s Two-Step CMOS A/D Converter," *IEEE J. Solid-State Circuits*, vol. SC-27, pp. 1667-1678, Dec. 1992.

5

Digital-to-Analog Converter Architectures

With the basic principles of D/A conversion explained in Chapter 4, we can now study this function from an architectural perspective. This chapter describes D/A converter architectures based on resistor ladders and current-steering arrays, with emphasis on stand-alone applications. While the capacitor DACs introduced in Chapter 4 are frequently used in A/D converters, they have not been popular as stand-alone circuits, primarily because they require a buffer to drive resistive loads and are also susceptible to nonlinear capacitive loading (e.g., due to switch junction capacitance).

5.1 RESISTOR-LADDER DAC ARCHITECTURES

The simplicity of resistor-ladder DACs using MOS switches makes these architectures attractive for many applications. However, for resolutions of 8 bits and above a simple ladder such as that described in Section 4.3.1 suffers from several drawbacks: it requires a large number of resistors and switches (2^m, where m is the resolution) and exhibits a long delay at the output. Consequently, alternative ladder topologies have been devised to improve the speed and resolution.

This section describes several resistor-ladder DAC architectures that provide means of overcoming the above problems.

5.1.1 Ladder Architecture with Switched Subdivider

In high-resolution applications, the number of devices in a DAC can be prohibitively large. It is therefore plausible to decompose the converter into a coarse section and a fine section so that the number of devices becomes proportional to approximately $2^{m/2}$ rather than 2^m, where m is the overall resolution. Such an architecture is shown in Figure 5.1(a). In this circuit, a primary ladder divides the main reference voltage, generating 2^j equal voltage segments. One of these segments is selected by the j most significant bits of the input and subdivided by a factor of 2^k using a secondary ladder such that $k + j = m$. If $k = j$, the number of devices in this architecture is proportional to $2^{m/2}$. It is also possible to utilize more than two ladders to further reduce the number of devices at high resolutions.

Figure 5.1(b) depicts a simple implementation of this architecture using MOS switches that are driven by 1-of-n codes in both stages [1]. Depending on the environment, these codes are generated from binary or thermometer code inputs.

The architecture of Figure 5.1(b) suffers from several drawbacks due to the switched subdivider. First, the finite on-resistance of the MOS switches used in the first multiplexer introduces differential nonlinearity at the output. Figure 5.2 illustrates this effect. Here, R_{u1} and R_{u2} represent the unit resistors of the primary and secondary ladders, respectively, and R_{on} denotes the switch on-resistance. Neglecting the current drawn by the secondary ladder, suppose the input code is such that the subdivider is connected to nodes $n - 1$ and n and the output voltage is equal to V_n [Figure 5.2(a)]. Now, if the digital input increments by 1 LSB, the subdivider switches to nodes n and $n + 1$ and, as shown in Figure 5.2(b), V_X is taken to the output. Ideally, the difference between V_X and V_n must be equal to 1 LSB $[= (V_{n+1} - V_n)/2^k]$. But, with the finite on-resistance of the switches, a finite error results. It can be easily shown that this error is approximately equal to $(V_{n+1} - V_n)R_{on}/(2R_{on} + 2^k R_{u2})$. In order to maintain this error much less than 1 LSB, R_{on} must be much less than R_{u2}.

Another source of differential nonlinearity in this architecture is the loading of the switched subdivider on the primary ladder. If the subdivider is connected to nodes $n - 1$ and n of the primary ladder (Figure 5.2), then $V_n - V_{n-1}$ is slightly less than $V_{n+1} - V_n$. It can be easily shown that for this error to be much less than 1 LSB, $R_{u1} \ll R_{u2}$.

Switched subdividers also exhibit long settling times. This occurs because when the subdivider switches from one segment of the primary ladder to another, all the capacitance associated with the subdivider (MOS device capacitance, wiring capacitance, etc.) must charge or discharge through the resistance of the primary ladder, a particularly acute problem if the digital input goes from zero to full-scale.

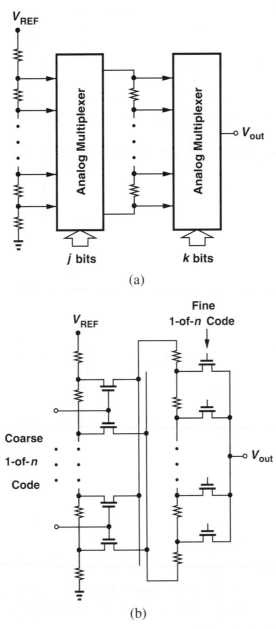

Fig. 5.1 Resistor-ladder DAC with switched subdivider. (a) Block diagram; (b) possible implementation.

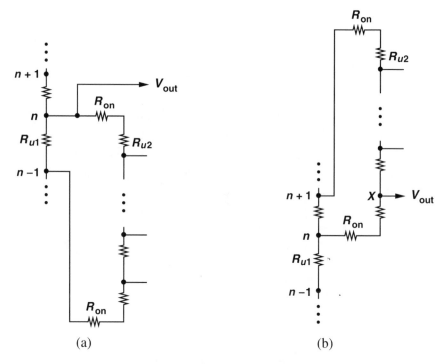

Fig. 5.2   Equivalent circuit of ladder DAC with switched subdivider. (a) Subdividing $V_n - V_{n-1}$; (b) subdividing $V_{n+1} - V_n$.

5.1.2 Intermeshed Ladder Architectures

Some of the drawbacks previously mentioned for ladder DACs can be alleviated through the use of intermeshed ladder architectures [2, 3]. In these architectures, a primary ladder divides the main reference voltage into equal segments, each of which is subdivided by a separate, fixed secondary ladder. Figure 5.3(a) illustrates such an arrangement [2], where all the switches are controlled by a 1-of-n code.

The intermeshed ladder has several advantages over single-ladder or switched-ladder architectures. Compared with a single-ladder DAC having the same resolution, this configuration can have smaller equivalent resistance at each tap, thus allowing faster recovery. Also, since the secondary ladders do not switch, their loading on the primary ladder is constant and uniform. Furthermore, the DNL resulting from finite on-resistance of switches does not exist here.

In the topology of Figure 5.3(a), the output node is loaded with the parasitic capacitance of all the switches connected to the ladder. To reduce

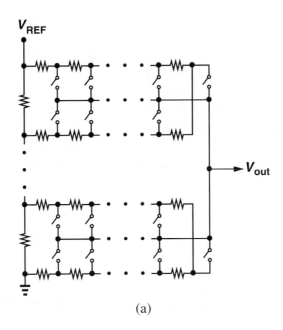

(a)

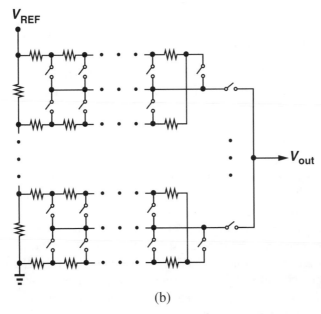

(b)

Fig. 5.3   Intermeshed resistor-ladder DACs with (a) one-level multiplexing and (b) two-level multiplexing.

this capacitance, the output nodes of the secondary ladders can be multiplexed [3] as depicted in Figure 5.3(b). In this circuit, the switches connected to the output node comprise a multiplexer and are controlled by the most significant bits. Note that in contrast with Figure 5.3(a), the resistive path charging the output node consists of *two* switches in series, but the overall settling time is less because the output node capacitance is reduced substantially.

Another approach to reducing the settling time is to precharge the DAC output node to a proper voltage [3]. For example, in Figure 5.3(b), the output node of each secondary ladder can be precharged to the middle tap voltage of that ladder so that the output voltage is already a coarse estimate of its final value.

5.2 CURRENT-STEERING ARCHITECTURES

Most high-speed D/A converters are based on current-steering architectures. Since these architectures can drive resistive loads directly, they do not require high-speed amplifiers at the output and hence are potentially faster than other types of DACs. While the high-speed switching of bipolar transistors makes them the natural choice for current-steering DACs, many designs have been recently reported in CMOS technology as well. Examples of high-speed design include 10-bit resolution at 1 GHz [4] and 12-bit resolution at 125 MHz [5].

5.2.1 *R*-2*R*-Network Based Architectures

In order to realize binary weighting in a current-steering DAC, an R-$2R$ ladder can be incorporated so as to relax device scaling requirements [6, 7]. Figure 5.4 illustrates an architecture that employs an R-$2R$ ladder in the emitter network. Here, the area of the bipolar transistors is scaled in powers of 2 from $2^{k-1}A_E$ for the MSB current source (Q_{k-1}) to A_E for the LSB current source (Q_0), where A_E denotes the emitter area of a unit transistor. Transistor Q_a, identical with Q_0, and its associated emitter resistors provide proper "termination" at the end of the array to allow accurate binary scaling. This point is clarified in the following analysis.

To see how binary weighting of currents is accomplished, we consider a 3-bit version of this architecture, shown in Figure 5.5(a), where mismatch effects are neglected. Since Q_a and Q_0 have equal emitter area and emitter resistors, their collector currents are equal, $I_a = I_0$. Consequently, as depicted in Figure 5.5(b), these transistors and their corresponding emitter resistors can be replaced with an equivalent circuit consisting of a transistor with emitter area $2A_E$ (Q_{0a}) and an emitter resistor equal to R. Now, Q_{0a} and Q_1 have equal emitter area and emitter resistors; i.e., $I_1 = I_0 + I_a = 2I_a$. Replacing

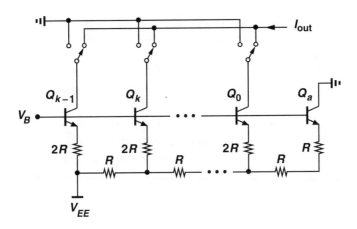

Fig. 5.4   Current-steering DAC with R-$2R$ ladder in the emitter network.

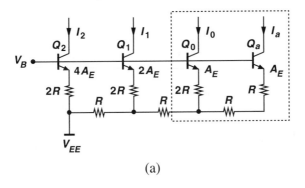

(a)

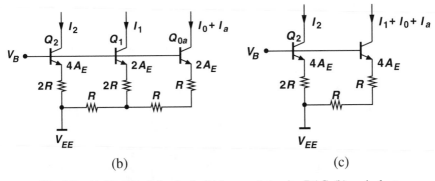

(b)   (c)

Fig. 5.5   (a) Simplified circuit of a 3-bit current-steering DAC; (b) equivalent circuit of (a); (c) equivalent circuit of (b).

these devices with an equivalent circuit yields that in Figure 5.5(c), from which it follows that $I_2 = I_1 + I_0 + I_a = 4I_a$. Thus, I_2, I_1, and I_0 are binary-weighted.

In contrast with the binary weighting discussed in Section 4.3.2 and depicted in Figure 4.8, the architecture of Figure 5.4 does not require a wide range of scaling for the resistors. Nonetheless, the transistors must still be scaled, a requirement that results in both large chip area and large capacitance at their collector nodes. To mitigate these problems, some of the LSB transistors can remain unscaled, with the resulting error corrected by additional circuit techniques [8]. As an example, consider the two current sources shown in Figure 5.6(a), where the emitter resistors of Q_1 and Q_2 are scaled by a factor of 2 but the transistors have equal area. If the voltage drop across the emitter resistors is much greater than V_T, then $I_2 \approx 2I_1$ and

$$V_{BE,2} - V_{BE,1} = V_T \ln \frac{I_2}{I_{S2}} - V_T \ln \frac{I_1}{I_{S1}} \tag{5.1}$$

$$\approx V_T \ln 2, \tag{5.2}$$

where I_{S1} and I_{S2} are the saturation currents of Q_1 and Q_2, respectively. Thus, I_1 is greater than its ideal value by approximately $(V_T \ln 2)/(2R)$.

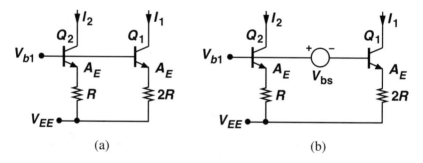

(a)   (b)

Fig. 5.6   (a) Current sources with scaled emitter resistors but unscaled transistors; (b) error correction using a voltage source equal to $V_T \ln 2$.

Since $I_2 \neq 2I_1$, (5.2) is merely an approximation, but its accuracy can be improved by iteration. For example, from the above approximation, $I_2 \approx 2I_1 + (V_T \ln 2)/R$, which can be substituted in (5.1):

$$V_{BE,2} - V_{BE,1} = V_T \ln \frac{2I_1 + (V_T \ln 2)/R}{I_1} \tag{5.3}$$

$$\approx V_T(1 + \frac{V_T}{2RI_1}) \ln 2. \tag{5.4}$$

Nonetheless, typically $2RI_1 \gg V_T$, and (5.2) provides a reasonable approximation.

The above discussion indicates that the base-emitter voltage of Q_1 in Figure 5.6(a) is $V_T \ln 2$ volts less than its ideal value for proper scaling. This error can be canceled by inserting a voltage source equal to $V_T \ln 2$ between the bias voltages applied to the bases of Q_1 and Q_2 [Figure 5.6(b)]. This voltage difference can be established by passing a current of $(V_T \ln 2)/R_{bs}$ through a resistor R_{bs} interposed between the bases of Q_1 and Q_2 [Figure 5.7(a)]. The current is proportional to absolute temperature (PTAT) and can be generated by any of the various band gap reference circuits [9, 10]. As an example, consider the circuit in Figure 5.7, where the emitter area of Q_2 is twice that of Q_1; i.e., $I_{S2} = 2I_{S1}$. The feedback loop established by the difference amplifier A ensures that $I_{C1} = I_{C2}$. Consequently,

$$V_{BE,2} - V_{BE,1} = V_T \ln \frac{I_{C2}}{I_{S2}} - V_T \ln \frac{I_{C1}}{I_{S1}} \tag{5.5}$$

$$= -V_T \ln 2. \tag{5.6}$$

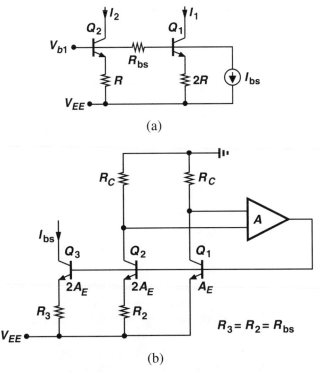

(a)

(b)

Fig. 5.7   (a) Implementation of a floating voltage of $R_{bs} I_{bs} = V_T \ln 2$; (b) generation of PTAT current $I_{bs} = (V_T \ln 2)/R_{bs}$.

This voltage difference is sustained across R_2 $(= R_{bs})$, giving $I_{C2} = (V_T \ln 2)/R_{bs}$. The output current I_{C3} is generated by a current source consisting of Q_3 and R_3, which are replicas of Q_2 and R_2, respectively. Note that the base current of Q_1 in Figure 5.7(a) introduces an additional voltage drop across R_{bs}, degrading the accuracy of correction slightly.

Another R-$2R$-network-based architecture is shown in Figure 5.8, where device scaling does not exceed a factor of 2 to 1, thereby reducing the circuit area substantially [7]. This circuit consists of identical transistors $Q_0, \ldots, Q_{k-1}$ with equal emitter resistors and an R-$2R$ network that performs the binary division of the collector currents.

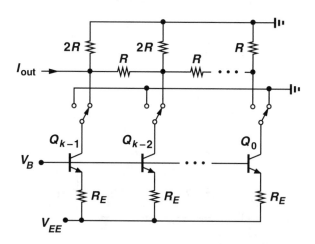

Fig. 5.8 Current-steering DAC with R-$2R$ ladder in the collector network.

To see how this division is accomplished, we consider a 3-bit example having the equivalent circuit shown in Figure 5.9(a) (with all the currents switched to the output). If the circuit inside the dashed box is replaced with its Norton equivalent, the circuit shown in Figure 5.9(b) is obtained. Combining the parallel devices in this circuit and substituting the result with another Norton equivalent, we arrive at Figure 5.9(c). Thus, the output current is a binary-weighted sum of I_1, I_2, and I_3. This derivation also shows that if an *external* resistor is tied from the output node to ground, it merely affects the full-scale output voltage swing and has no influence on the accuracy of the binary weighting.

The architectures described in this section can be employed to achieve high resolutions. The small number of resistors—twice the number of bits—

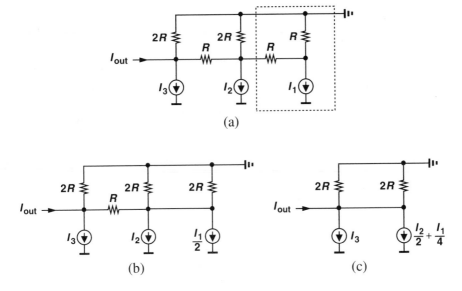

Fig. 5.9 Equivalent circuits of a 3-bit current-steering DAC with an R-$2R$ ladder in the collector network.

used in these circuits makes it possible to laser-trim these resistors, thus correcting mismatch errors. However, unlike segmented arrays, these architectures do not exploit error averaging due to a large number of nominally identical devices and hence require tighter matching of the components.

Another drawback of these architectures is their potentially large glitch area. To study this effect, consider the simplified circuit of a 4-bit DAC shown in Figure 5.10. First, suppose the digital binary input is equal to 1000 and hence only the MSB current source is switched to the output. Now, if the digital input goes to 0111, the MSB current source turns off while the other three turn on and the output changes by IR_L. In practice, however, the digital signals driving the switches suffer from finite risetime and falltime as well as timing skews. For example, during the transition from 1000 to 0111, all four switches may be partially off for a short time. Thus, the output current momentarily reaches a value different from either $8I$ or $7I$, causing a glitch.

Timing skews can be suppressed through the use of on-chip latches to sample and align the incoming digital signals and apply only the sampled values to the current switches. Finite transition times and clock skews, however, still result in output glitches.

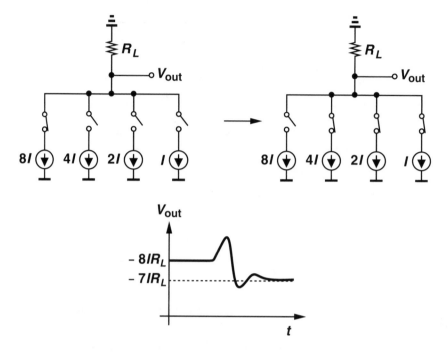

Fig. 5.10   Glitch impulse in a current-steering DAC.

5.2.2 Segmented Architectures

As mentioned in the previous section, architectures based on simple binary weighting suffer from two important drawbacks. First, they require tight device matching to achieve monotonicity (DNL < 1 LSB). Second, they exhibit large glitch impulses. Segmented architectures are commonly used to alleviate these problems.

Figure 5.11 shows an m-bit segmented current-steering architecture which is based on the segmented array described in Section 4.3.2. In this architecture, $N = 2^m - 1$ nominally identical current sources are controlled by a thermometer code. In a stand-alone DAC, a binary-thermometer encoder precedes the array.

In Figure 5.11, as the digital input increases, the DAC switches more current sources to the output without turning off any of the current sources that are already switched to the output. Thus, the input/output characteristic is monotonic and the glitch impulse is small. Note that while monotonicity typically results in a small DNL, the INL may be quite large if the devices do not match accurately.

The circuit of Figure 5.11 is called a "fully segmented" architecture because it employs $2^m - 1$ equal current sources for m bits. For $k \geq 8$, the

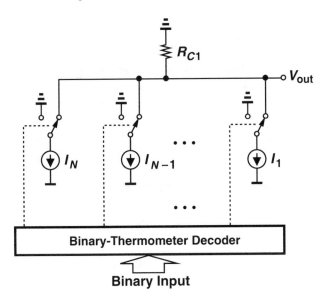

Fig. 5.11   Segmented current-steering DAC.

number of devices in the array becomes quite large, leading to a high capacitance at the output node. Furthermore, the binary-thermometer decoding logic occupies a large area and requires substantial power dissipation. Even though fully segmented architectures have been used for resolutions as high as 10 bits [11] (to minimize the glitch impulse), it is often more efficient to partition the DAC into a segmented coarse sub-DAC and a binary-weighted fine sub-DAC whose output currents are simply added. In other words, for a resolution of $m = k + n$, the k most significant bits are converted to thermometer code and drive a 2^k-unit segmented array, while the remaining n bits are directly applied to an n-bit binary array. Called the partially segmented architecture, this topology is illustrated for $k = 6$ and $n = 4$ in Figure 5.12. Here, an R-$2R$ ladder performs the binary weighting in the same manner as described in Section 5.2.1.

The choice of k and n in general depends on the matching of the current sources and the tolerable glitch area. Typical values for k range from 4 to 7.

In the architecture of Figure 5.12, the complexity of the binary-thermometer code conversion, the large number of current sources, and the issues related to routing signals place stringent requirements on floor planning and layout. An efficient approach is to arrange the current sources and part of the decoding in a matrix, as shown in Figure 5.13(a) [12]. Here, binary-thermometer conversion is performed in two steps: row and column decoding followed by local decoding within each cell of the matrix. In the first step, the

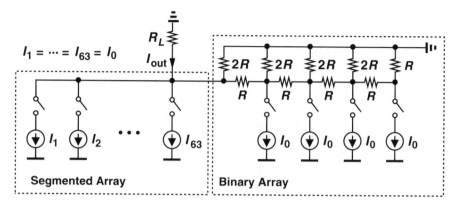

Fig. 5.12   A 10-bit partially segmented current-steering DAC.

input binary word ($D_6 \ldots D_1$) is partitioned into two subwords ($D_6 D_5 D_4$ and $D_3 D_2 D_1$) each of which is converted to a thermometer code. The resulting codes are distributed across the matrix as depicted in Figure 5.13(a). In the second step, the row and column thermometer codes are combined locally to determine the required status of each current source. To implement local decoding, we note that for any digital input, each row falls in one of three categories [12]: (1) rows in which all current cells are on, (2) rows in which all current cells are off, and (3) a certain row in which some of the current cells are on. To determine the status of each cell, the row and column thermometer codes are locally combined around each current source. Figure 5.13(b) shows an example where two adjacent bits of the row thermometer code and one bit of the column thermometer code are used to generate the control signal for the current source.

While providing a simple, modular layout, the matrix configuration must deal with two difficulties. First, the analog output line inevitably experiences a great deal of coupling from the digital signals that flow in both horizontal and vertical directions within the matrix. In bipolar implementations, this effect is minimized through the use of relatively small (≈ 0.5 V) differential signals for column and row thermometer codes. In CMOS circuits, on the other hand, these signals are typically rail-to-rail and single-ended, resulting in substantial coupling to the analog output.

The second issue stems from the wiring capacitance of the analog output line. This line must reach all the cells, and its total length is given primarily by the size of each cell. Since the cells must be large enough to accommodate local decoding (and a current source), the total length of the analog output line and hence the output settling are quite long.

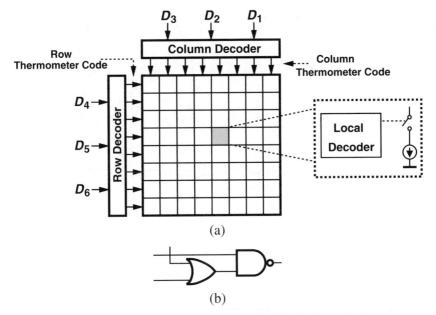

Fig. 5.13   (a) Matrix floorplan for segmented DACs, (b) local decoding.

The idea of segmented DACs was originally proposed by Shoeff and used in a 12-bit converter to achieve monotonicity without trimming [13]. Shown in Figure 5.14 is a simplified 5-bit version of this architecture consisting of a segmented array I_4-I_7 and a binary-weighted array I_0-I_3. The basic principle of this architecture is to switch each of the equal current sources I_4-I_7 to the output in an additive fashion and use their subdivisions to provide finer steps and hence higher resolution.

To illustrate this principle, we examine the output current as the digital (binary) input $D_{in} = D_5 \cdots D_1$ varies from zero to full-scale. For $0 \leq D_{in} \leq 2^3 - 1$, the current source I_4 is switched to node P and subdivided to generate I_3-I_0. Thus, I_3-I_1 are switched to the output node according to the value of D_{in}, while other currents are switched to ground. Note that for $D_{in} = 2^3 - 1$, $I_{out} = I_4 - I_0$.

When the digital input is equal to 2^3, I_4 is switched to the output node, I_5 to node P, and the remaining currents to ground. Thus, the difference between the output currents corresponding to $D_{in} = 2^3 - 1$ and $D_{in} = 2^3$ is equal to I_0, indicating that this transition is always monotonic because I_0 has a finite positive value.

For $2^3 \leq D_{in} \leq 2^4 - 1$, I_5 is subdivided to generate I_3-I_0, and the output current is equal to the sum of a proper combination of these currents and I_4.

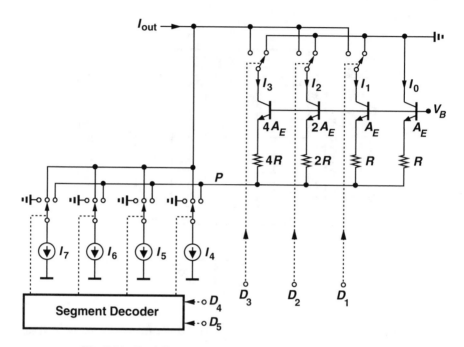

Fig. 5.14   Partially segmented architecture proposed by Shoeff.

For $D_{in} = 2^4$, I_5 is switched directly to the output; i.e., $I_{out} = I_4 + I_5$. Again, the difference between output currents corresponding to $D_{in} = 2^4 - 1$ and $D_{in} = 2^4$ is equal to I_0, thereby preserving monotonicity. The same switching algorithm is used for larger digital inputs.

The integral linearity of this architecture depends primarily on the matching of I_4-I_7. Laser-trimming or self-calibration techniques such as those described in Section 8.3.1 can be employed to achieve a small INL.

Despite the above features, this circuit faces an important limitation when implemented in today's technologies. The stacking of devices in current sources I_3-I_0 on top of one of I_4-I_7 severely limits the output voltage swing if the supply voltage is only 5 V.

The segment decoder utilized in this architecture is more than a binary-to-thermometer converter because it must also determine which one of I_4-I_7 is switched to node P. A compact implementation is possible if multiple logic levels are used [13].

REFERENCES

[1] H. U. Post and K. Schoppe, "A 14 Bit Monolithic NMOS D/A Converter," *IEEE J. Solid-State Circuits*, vol. SC-18, pp. 297-302, June 1983.

[2] A. G. Dingwall and V. Zazzu, "An 8-MHz CMOS Subranging 8-Bit A/D Converter," *IEEE J. Solid-State Circuits*, vol. SC-20, pp. 1138-1143, Dec. 1985.

[3] M. J. M. Pelgrom, "A 10-Bit 50-MHz CMOS D/A Converter with 75-Ω Buffer," *IEEE J. Solid-State Circuits*, vol. SC-25, pp. 1347-1352, Dec. 1990.

[4] P. Vorenkamp et al., "A 1 Gs/s 10 b Digital-to-Analog Converter," *ISSCC Dig. Tech. Pap.*, pp. 52-53, Feb. 1994.

[5] W. T. Sagun et al., "A 125-MHz 12-Bit Digital-to-Analog Converter System," *Hewlett-Packard J.*, pp. 78-85, April 1988.

[6] G. Kelson, H. H. Stellrecht, and D. S. Perloff, "A Monolithic 10-b Digital-to-Analog Converter Using Ion Implantation," *IEEE J. Solid-State Circuits*, vol. SC-8, pp. 396-403, Dec. 1973.

[7] D. J. Dooley, "A Complete Monolithic 10-b D/A Converter," *IEEE J. Solid-State Circuits*, vol. SC-8, pp. 404-408, Dec. 1973.

[8] P. Holloway and M. Norton, "A High Yield, Second Generation 10-Bit Monolithic DAC," *ISSCC Dig. Tech. Pap.*, pp. 106-107, Feb. 1976.

[9] A. P. Brokaw, "A Simple Three-Terminal IC Bandgap Reference," *IEEE J. Solid-State Circuits*, vol. SC-9, pp. 388-393, Dec. 1974.

[10] R. J. Widlar, "New Developments in IC Voltage Regulators," *IEEE J. Solid-State Circuits*, vol. SC-6, pp. 2-7, Feb. 1971.

[11] H. Takakura and M. Yokoyama, "A 10 Bit 80 MHz Glitchless CMOS D/A Converter," *Proc. CICC*, pp. 26.5.1-26.5.3, May 1991.

[12] T. Miki, "An 80-MHz 8-Bit CMOS D/A Converter," *IEEE J. Solid-State Circuits*, vol. SC-21, pp. 983-988, Dec. 1986.

[13] J. A. Shoeff, "An Inherently Monotonic 12 Bit DAC," *IEEE J. Solid-State Circuits*, vol. SC-14, pp. 904-911, Dec. 1979.

6

Analog-to-Digital Converter Architectures

Analog-to-digital converters provide the link between the analog world and digital systems. Due to their extensive use of analog and mixed analog-digital operations, A/D converters often appear as the bottleneck in data processing applications, limiting the overall speed or precision.

Following a definition of performance metrics, we describe in this chapter a number of A/D converter architectures commonly employed in high-performance systems. These architectures can be broadly classified as one-step or multistep, with flash, interpolative, and folding topologies in the first category and two-step, pipelined, and successive approximation configurations in the second. However, two-step ADCs (and the concept of residue) are discussed before interpolative and folding architectures to simplify the description of the latter. Also, flash and two-step converters are studied in great detail because a good understanding of these two architectures proves helpful in analyzing other ADCs as well.

In addition to the ADC architectures covered here, several others are sometimes used in specialized applications. Gordon provides a comprehensive study of such architectures [1].

6.1 GENERAL CONSIDERATIONS

An A/D converter produces a digital output, D, as a function of the analog input, A:

$$D = f(A) \qquad (6.1)$$

While the input can assume an infinite number of values, the output can be selected from only a finite set of codes given by the converter's output word length (i.e, resolution). Thus, the ADC must approximate each input level with one of these codes. This is accomplished, for example, by generating a set of reference voltages corresponding to each code, comparing the analog input with each reference, and selecting the reference (and its code) closest to the input level. In most ADCs, the analog input is a voltage quantity because comparing, routing, and storing are easier for voltages than for currents.

Figure 6.1(a) depicts a simple ADC input/output characteristic where the analog input is approximated with the *nearest smaller* reference level. If the digital output is an *m*-bit binary number, then

$$D = [2^m \frac{A}{V_{REF}}], \tag{6.2}$$

where [·] denotes the integer part of the argument and V_{REF} is the input full-scale voltage. Note that the minimum change in the input that causes a change in the output is $\Delta = V_{REF}/2^m$ and corresponds to the least significant bit of the digital representation.

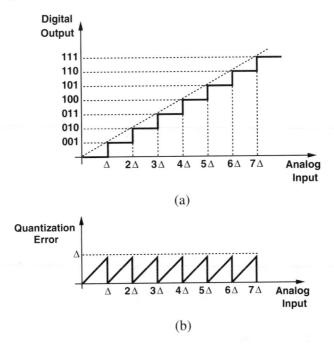

(a)

(b)

Fig. 6.1   (a) Input/output characteristic; (b) quantization error of an A/D converter.

The approximation or "rounding" effect in A/D converters is called "quantization," and the difference between the original input and the digitized output is called the "quantization error" and is denoted here by ε_q. For the characteristic of Figure 6.1(a), ε_q varies as shown in Figure 6.1(b), with the maximum occurring before each code transition. This error decreases as the resolution increases, and its effect can be viewed as additive noise (called "quantization noise") appearing at the output. Thus, even an "ideal" m-bit ADC introduces nonzero noise in the converted signal simply due to quantization.

We can formulate the impact of quantization noise on the performance as follows. For simplicity, consider a slightly different input/output characteristic, shown in Figure 6.2(a), where code transitions occur at odd (rather than even) multiples of $\Delta/2$. A time domain waveform therefore experiences both negative and positive quantization errors as illustrated in Figure 6.2(b). To calculate the power of the resulting noise, we assume that ε_q is (1) a random variable uniformly distributed between $-\Delta/2$ and $+\Delta/2$, and (2) independent of the analog input. While these assumptions are not strictly valid in the general case, they usually provide a reasonable approximation for resolutions above 4 bits. The quantization noise power can then be expressed as the mean square of ε_q:

$$\overline{\varepsilon_q^2} = \frac{1}{\Delta} \int_{-\Delta/2}^{+\Delta/2} \varepsilon_q^2 \, d\varepsilon_q \tag{6.3}$$

$$= \frac{\Delta^2}{12}. \tag{6.4}$$

If the analog input is a sinusoid with amplitude $V_{\text{REF}}/2$ (peak-to-peak V_{REF}), its total power is equal to $V_{\text{REF}}^2/8 = 2^{2m}\Delta^2/8$. Thus, the peak signal-to-noise ratio SNR_P at the output is

$$\text{SNR}_P = \frac{2^{2m-3}\Delta^2}{\Delta^2/12} \tag{6.5}$$

$$= \frac{3}{2}2^{2m}, \tag{6.6}$$

which, when expressed in decibels, becomes

$$\text{SNR}_P = 6.02m + 1.76 \quad \text{dB}. \tag{6.7}$$

This equation is often used to compare the performance of a given m-bit ADC with that of an ideal one. For example, the peak SNR may be measured independently to attain an effective value for m. This is discussed in more detail later.

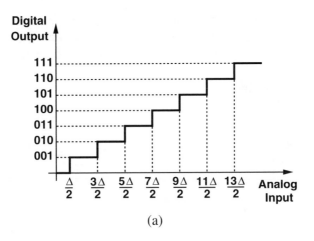

(a)

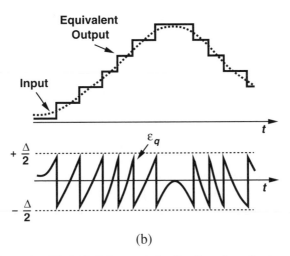

(b)

Fig. 6.2   (a) Modified ADC characteristic; (b) effect of amplitude quantization on a time domain waveform.

The quantization characteristics described above are "uniform"; i.e., their transition points are equally spaced. In some applications, other types such as logarithmic quantization are desirable [2]. In this book, we consider only uniform quantization.

A more extensive analysis of quantization can be found in [2] and [3].

6.2 PERFORMANCE METRICS

As with sampling circuits and D/A converters, full specification of the performance of ADCs requires a large number of parameters, some of which are defined differently by different manufacturers. Here, we define a number of important metrics frequently used in the book. For a complete set of specifications, the reader is referred to the literature [4, 5] and manufacturers' data books.

Illustrated in Figure 6.3, the following definitions describe the static behavior of ADCs.

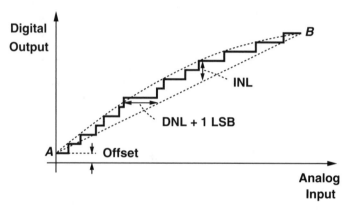

Fig. 6.3   Static ADC metrics.

- Differential nonlinearity (DNL) is the maximum deviation in the difference between two consecutive code transition points on the input axis from the ideal value of 1 LSB.

- Integral nonlinearity (INL) is the maximum deviation of the input/output characteristic from a straight line passed through its end points (line AB in Figure 6.3). The overall difference plot is called the INL profile.

- Offset is the vertical intercept of the straight line through the end points.

- Gain error is the deviation of the slope of line AB from its ideal value (usually unity).

Often specified as a function of the sampling and input frequencies, the following terms are used to characterize the dynamic performance of converters.

- Signal-to-noise ratio (SNR) is the ratio of the signal power to the total noise power at the output (usually measured for a sinusoidal input).

- Signal-to-(noise + distortion) ratio (SNDR) is the ratio of the signal power to the total noise and harmonic power at the output, when the input is a sinusoid.

- Effective number of bits (ENOB) is defined by the following equation:

$$\text{ENOB} = \frac{\text{SNDR}_P - 1.76}{6.02}, \qquad (6.8)$$

where SNDR_P is the peak SNDR of the converter expressed in decibels.

- Dynamic range is the ratio of the power of a full-scale sinusoidal input to the power of a sinusoidal input for which SNR= 0 dB.

It is important to note that since A/D conversion entails signal sampling, many of the sampling circuit metrics described in Chapter 2 can be added to ADC parameters. For example, aperture jitter is a crucial parameter because it directly impacts the output SNR.

6.3 FLASH ARCHITECTURES

Conceptually the simplest and potentially the fastest, flash architectures employ parallelism and "distributed" sampling to achieve a high conversion speed. Figure 6.4 is a block diagram of an m-bit flash ADC. The circuit consists of 2^m comparators, a resistor ladder comprising 2^m equal segments, and a decoder. The ladder subdivides the main reference into 2^m equally spaced voltages, and the comparators compare the input signal with these voltages. For example, if the analog input is between V_j and V_{j+1}, comparators A_1 through A_j produce ONEs at their outputs while the rest generate ZEROs. Consequently, the comparator outputs constitute a thermometer code (Chapter 4), which is converted to binary by the decoder.

As explained in Chapter 7, comparators often incorporate clocked regenerative amplifiers to achieve a high speed. In a flash architecture, clocked comparators offer another advantage as well: they act as "polarity sampling" circuits. We illustrate this by considering a simple block diagram of a comparator (Figure 6.5). This circuit consists of a preamplifier A and a latch. When Φ is high, A amplifies the difference between $V_{\text{in},1}$ and $V_{\text{in},2}$ while the latch is disabled. When Φ goes low and $\overline{\Phi}$ goes high, A is disabled and the latch begins to amplify the difference established at its input by the preamplifier, thus generating logic levels at the output. In other words, when Φ is high, the comparator tracks the input, and when Φ goes low, it stores the instantaneous polarity of $V_{\text{in},1} - V_{\text{in},2}$. Thus, if all the comparators in Figure 6.4 are strobed at the same time, they collectively store the polarity of the difference between V_{in} and each V_j, thereby operating as a "distributed" sample-and-hold circuit.

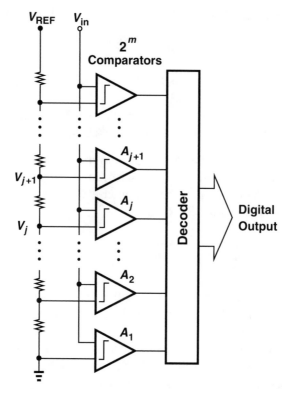

Fig. 6.4 Block diagram of an m-bit flash A/D converter.

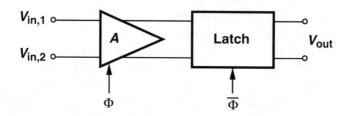

Fig. 6.5 Simple comparator architecture.

It follows from the above discussion that full-flash architectures, in principle, do not need explicit front-end sample-and-hold circuits and their performance is determined primarily by that of their constituent comparators. Since comparators do not require linear amplification (Chapter 7) and hence typically achieve much higher speeds than sample-and-hold amplifiers, flash ADCs can operate faster than those that demand front-end SHAs.

Despite these features, flash topologies suffer from a number of drawbacks due to massive parallelism or lack of a front-end sampling circuit. Since the number of comparators grows exponentially with the resolution, these ADCs require excessively large power and area for resolutions above 8 bits. Furthermore, the large number of comparators gives rise to problems such as dc and ac deviation of the reference voltages generated by the ladder, large *nonlinear* input capacitance, and kickback noise at the analog input. In addition, the lack of a front-end sample-and-hold amplifier makes the converter susceptible to sparkles and slew-dependent sampling points. We discuss each of these issues below.

6.3.1 Reference Ladder DC and AC Bowing

The nonidealities of the comparators in a flash ADC can introduce substantial errors in the reference voltages produced by the resistor ladder. The input bias current of bipolar comparators gives rise to dc "bowing" of these voltages, while capacitive feedthrough or switched capacitances at the input of each comparator cause transient errors. Both of these errors can also be viewed as integral nonlinearity.

To illustrate the dc bowing, let us consider the bipolar stage shown in Figure 6.6, a circuit often used at the front end of comparators. Here, the emitter follower Q_2 draws a finite, relatively constant base current, I_B, from the reference ladder, thereby lowering each tap voltage. The total error due to all the comparators can be obtained using the equivalent circuit of Figure 6.7(a), where R_u denotes the ladder unit resistance and both ends of the ladder are grounded so that each node voltage contains only an error term. To find the voltage at node j, we replace the circuits in the dashed boxes with their Norton equivalents. Using superposition, the reader can easily prove that the circuit in Figure 6.7(b) is attained. The integral nonlinearity at node j is therefore equal to

$$\text{INL}_j = \frac{1}{2} j (N - j) R_u I_B, \tag{6.9}$$

which, at $j = N/2$, reaches a maximum of

$$\text{INL}_{\text{max}} = \frac{1}{8} N^2 R_u I_B. \tag{6.10}$$

Note that this error increases with the *square* of the number of the comparators.

In order to reduce dc bowing, currents with proper values can be injected at one or more (equally spaced) points along the ladder. For example, a current

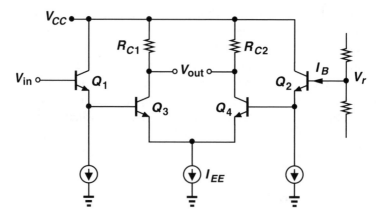

Fig. 6.6   Input stage of a typical bipolar comparator.

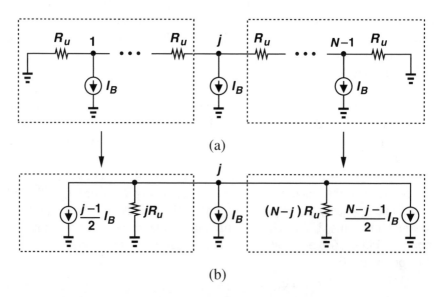

Fig. 6.7   (a) Equivalent circuit of a resistor ladder with dc bowing; (b) simplified model of (a) using Norton equivalent.

of $NI_B/2$ injected at $j = N/2$ removes the error at this node, lowering the maximum INL by a factor of 4. Such corrective currents must of course track the actual value of I_B so as to maintain accuracy with process and temperature variations.

Another error in the reference levels is caused by capacitive feedthrough from the analog input to the resistor ladder. Consider the bipolar stage shown in Figure 6.8, where the base-emitter capacitance of Q_1-Q_4 provides a signal path from V_{in} to V_r. For high-frequency inputs, this path conducts appreciably, thereby disturbing the dc voltage generated by the ladder at V_r. If the number of comparators is large, the total signal feedthrough becomes substantial and the ladder tap voltages deviate from their dc values significantly. Note that this error is also proportional to the square of the number of comparators.

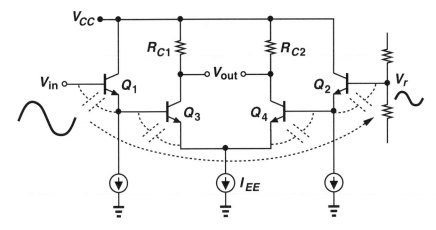

Fig. 6.8   Analog input feedthrough to reference ladder.

The transient error due to feedthrough can be alleviated by lowering the unit resistance of the ladder, but at the cost of higher power dissipation and difficulties in routing the ladder.

Another type of transient bowing arises in CMOS flash converters wherein comparators employ switched capacitors at their input. As explained in Chapter 8, CMOS comparators often incorporate offset cancellation for resolutions of 8 bits and above. Illustrated in Figure 6.9 is a typical case, where a CMOS inverter operates as a comparator [6, 10]. In the sampling mode, S_1 and S_3 are on and S_2 is off. Thus, the inverter is driven to its high-gain region and V_{in} is sampled on C_1. During comparison, only S_2 is on and the inverter generates an output voltage proportional to the difference between the sampled value of V_{in} and V_r. When C_1 is connected to V_r, it is in series with the input capacitance of the inverter. Since C_1 is usually much larger than this capacitance, we can reduce the circuit to an inverter input capacitance that switches between V_{in} and V_r every cycle. This in effect is equivalent to a resistive path between V_{in} and V_r, introducing ac bowing along the ladder. In practice, the parasitic capacitances of C_1 to the substrate aggravate this problem.

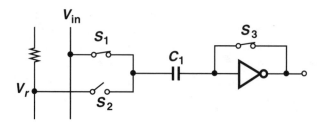

Fig. 6.9   Section of a CMOS flash ADC.

6.3.2 Nonlinear Input Capacitance

While the total input capacitance of a flash ADC (together with the signal source output impedance) limits the analog input bandwidth, nonlinearities in this capacitance introduce harmonic distortion in the sampled signal. For example, in a bipolar converter, each comparator exhibits a nonlinear input capacitance due to the base-collector junction of the input device (Figure 6.10). For a large number of comparators, this nonlinearity becomes substantial, and the low-pass filter formed by the signal source resistance and the ADC input capacitance yields an input-dependent delay. In other words, the input experiences a smaller delay when it is most negative than when it is most positive.

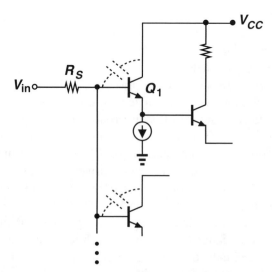

Fig. 6.10   Nonlinear input capacitance of a bipolar flash ADC.

To understand this effect, consider the simplified equivalent circuit shown in Figure 6.11, where R_S represents the signal source resistance and C_{in} is the total base-collector capacitance seen at the analog input. For $V_{in} = A \sin \omega t$ and *linear* C_{in}, the output would be $V_{out} \approx A \sin(\omega t + \theta)$, where amplitude attenuation is neglected and $\theta = -\tan^{-1}(R_S C_{in} \omega)$. The capacitance nonlinearity can be included in this equation by assuming that the output waveform closely resembles a sinusoid but exhibits a voltage-dependent phase shift. Since for small θ, $\tan^{-1} \theta \approx \sin \theta \approx \theta$ and

$$V_{out} \approx A \sin \omega t + A \theta \cos \omega t, \tag{6.11}$$

we have

$$V_{out} \approx A \sin \omega t - A R_S C_{in} \omega \cos \omega t. \tag{6.12}$$

Because C_{in} is a function of the input voltage, the second term in (6.12) contains sinusoidal components at 2ω, 3ω, etc. These components can be easily found by approximating C_{in} with a polynomial [such as that in equation (4.40)] and expanding the second term in (6.12).

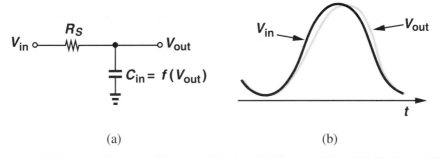

(a)   (b)

Fig. 6.11   (a) Simplified equivalent input circuit of a flash ADC; (b) waveforms of (a).

6.3.3 Kickback Noise

An important issue in flash converters is the accumulative effect of the kickback noise introduced by the comparators. As explained in Chapter 7, strobed comparators generate a great deal of noise at their inputs during the transition from latching to tracking. In the flash architecture of Figure 6.4, this noise sees different impedances at the input terminal and at the ladder taps, and hence produces a differential error. In high-speed applications, if this noise does not decay to sufficiently small levels from one cycle to the next, the analog input at the sampling instant is corrupted. The error resulting from this noise is proportional to the square of the number of comparators.

6.3.4 Sparkles in Thermometer Code

An important effect resulting from the lack of a SHA in flash ADCs is the so-called sparkles (or bubbles) in the thermometer code [13]. The mechanism leading to this effect is illustrated in Figure 6.12, where a fast-varying signal is applied to a flash converter. Suppose comparators A_j and A_{j+1} are not exactly identical. For example, suppose the actual strobing instant of A_j is slightly *earlier* than that of A_{j+1}; i.e., A_j switches from tracking to latching at t_1, while A_{j+1} does so at t_2. If the analog input varies by more than $V_{j+1} - V_j$ between t_1 and t_2, A_j senses a level *less* than V_j at t_1 and A_{j+1} a level *greater* than V_{j+1} at t_2. As a consequence, A_j generates a ZERO at its output, while A_{j+1} produces a ONE. Called a "sparkle", the out-of-sequence ZERO at the output of A_j not only degrades the signal-to-noise ratio of the sampled waveform but, more importantly, presents potential problems in subsequent thermometer-binary decoding as well. The SNR degrades because the thermometer code is not an accurate representative of the sampled waveform whereas decoding becomes more difficult because some of the common decoding topologies produce gross errors in the presence of sparkles.

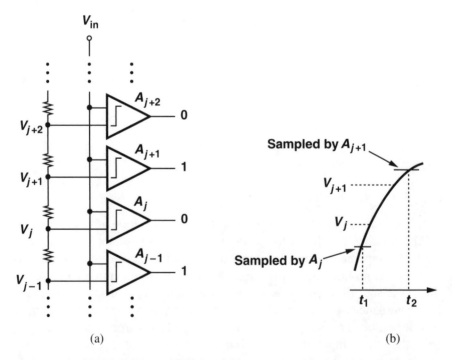

(a)   (b)

Fig. 6.12 Generation of sparkles in the thermometer code.

As an example, consider the thermometer-binary decoder in Figure 6.13, where a 1-of-n decoder is followed by a ROM. The 1-of-n decoder consists of two-input NAND gates that sense every two consecutive bits of the thermometer code, generating a ZERO if both are ZERO or both are ONE, and a ONE if the upper bit is ZERO and the lower bit is ONE. In the presence of a sparkle, two of the NAND gates produce ONEs at their outputs, activating two rows of the ROM simultaneously. In a standard ROM structure, this may result in a large error in the binary output. For example, if the two rows correspond to binary outputs 0111 and 1001, then the final output will be 1111, approximately one-half of full-scale away from the true value.

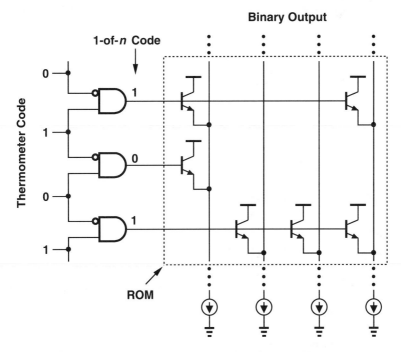

Fig. 6.13   Thermometer-binary decoder with a sparkle error.

The sparkle illustrated in Figure 6.12 is but a simple example. In practice, for high slew rate inputs several consecutive comparators may exhibit mismatches in their sampling instants such that multiple sparkles occur in the thermometer code.

The above discussion also indicates that if the *dc offset* of comparators A_j and A_{j+1} in Figure 6.12(a) is sufficiently large, then a similar error can occur. Nonetheless, comparators are usually designed so that their dc offset is much less than 1 LSB. As a result, sparkles are predominantly caused by timing mismatch among comparators.

Various circuit techniques have been devised to suppress the effect of sparkles [13, 14]. In a simple approach, the decoding that follows the thermometer code senses more than two consecutive bits of this code so as to detect sparkles. Shown in Figure 6.14 is an example where three-input NAND gates perform a first-order correction, generating a true 1-of-n code if the thermometer code has only one sparkle. More sophisticated methods are described in [13, 14, 15].

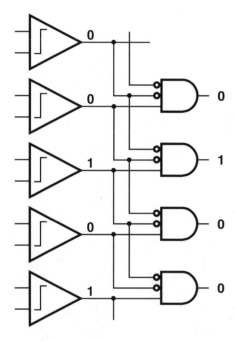

Fig. 6.14   Correction of sparkle in a thermometer code.

6.3.5 Metastability

Since flash architectures employ comparators, they are susceptible to metastability errors [16]. As explained in Chapter 7, metastability occurs when the difference at the input of a comparator is small, making the circuit take a long time to produce a well-defined logic output. If the instantaneous value of the input signal to a flash ADC is close to the reference voltage of one of the comparators, that comparator will have an indeterminate output for a long time, possibly causing an erroneous digital output for that particular conversion.

As an example, consider the decoding circuit in Figure 6.15, where comparator A_j is metastable. NAND gates G_j and G_{j+1} sense the output of this comparator and are to provide logic levels for two rows of the ROM, corresponding to binary outputs 0111 and 1000, respectively. If the output of A_j is indeterminate, it is possible that G_j interprets it as ZERO while G_{j+1} interprets it as ONE, thus causing the ROM to generate a binary output of 1111 and hence a gross error in the digital output.

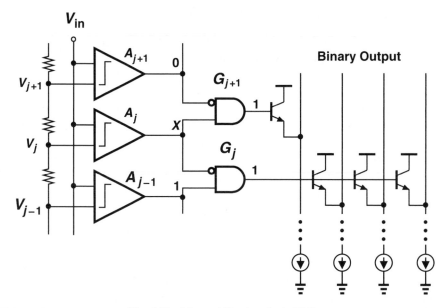

Fig. 6.15   Metastability in a flash ADC.

Note that when metastability occurs, the final logical value of the metastable comparator output is not critical; i.e., it can be either ZERO or ONE because the difference between the analog input and the comparator reference voltage is very small. However, it is the tardiness of the comparator in reaching a logical value that causes substantial errors.

From the example in Figure 6.15, we can see that if a metastable level is applied to more than one gate at the same time (i.e., if it "splits"), it can be interpreted differently by different gates and hence introduce large errors, but so long as it is sensed by only one input, it is likely to approach correct logic levels.

In order to lower the probability of metastable states, the thermometer-binary decoding can be pipelined so that potentially indeterminate outputs are allowed more regeneration time. This is of course possible only if each potentially metastable level is sensed by no more than one input, an unsurmountable issue in the decoder of Figure 6.15. A simple but power-consuming approach would be to follow each comparator with more latches to allow a longer regeneration time for the thermometer code. A more efficient technique employs Gray coding as an intermediate step between the thermometer and binary codes. This is discussed in Section 6.3.8.

6.3.6 Slew-Dependent Sampling Point

Another type of error in the presence of fast-varying analog inputs is slew-dependent sampling instant of comparators [12]. This error is due to the finite switching time from tracking to latching. We illustrate this effect using Figure 6.16, where a simple comparator is shown along with its waveforms. At $t = t_1$, CK goes high to switch the circuit from tracking to latching. As the latch turns on, it begins to amplify the initial differential voltage at nodes X and Y regeneratively, while the input differential pair begins to turn off. From the time the latch begins to turn on until the time the input pair turns off completely, the input signal can still influence V_{XY} through Q_1 and Q_2. For example, as depicted in Figure 6.16(b), if the input signal varies slowly, although the polarity of $V_{in} - V_r$ changes after $t = t_1$, the effect is not strong enough to override the polarity of V_{XY}. On the other hand, as shown in Figure 6.16(c), if $V_{in} - V_r$ changes rapidly, it can reverse the polarity of V_{XY} and generate a different logic output. This phenomenon can be viewed as a variation in the sampling instant of the comparator as a function of the input slew rate and introduces odd harmonics because it occurs for both negative and positive slopes [12].

The above error can be lowered by making the clock transition rates sufficiently higher than the maximum slew rate of the analog input. On a large chip, this requires careful distribution and buffering of the clock(s) with particular attention to their loading.

6.3.7 Clock Jitter and Dispersion

As explained in Chapter 2, all sampling circuits suffer from SNR degradation as the jitter of the sampling command increases, a particularly critical issue when the analog input has high slew rates. Flash (and other types of) ADCs are no exception because they incorporate sampling.

To arrive at a simple relation between maximum tolerable jitter and an ADC's speed and resolution, we can say that jitter has negligible effect on the overall SNR if the analog input varies by less than 1 LSB during jitter-induced time deviation of the sampling point. Thus, for a full-scale analog

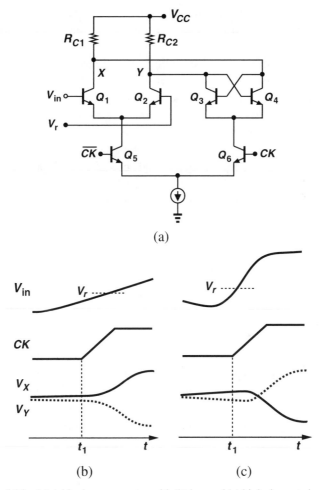

(a)

(b)   (c)

Fig. 6.16   (a) A bipolar comparator with (b) low and (c) high slew rate inputs.

input $V_{in} = A \sin 2\pi f t$, whose maximum rate of change is equal to $2\pi f A$, the above condition can be expressed as

$$2\pi f \, \Delta t \; < \; 1 \, \text{LSB} \qquad (6.13)$$

$$< \; \frac{2A}{2^m}, \qquad (6.14)$$

where Δt represents the clock jitter and m is the converter's resolution. It follows that

$$\Delta t \; < \; \frac{1}{\pi f 2^m}. \qquad (6.15)$$

In Nyquist-rate converters, f approaches half of the ADC conversion speed. This result is only a rough estimate, but it allows a quick calculation of jitter requirements.

The distributed nature of sampling in flash converters gives rise to a unique timing problem that does not exist in architectures with a single front-end sample-and-hold amplifier. Since the analog signal and the clock must travel long distances on a large ADC chip, they experience different delays due to different loading [16]. Furthermore, even with identical loading, the clock waveform—ideally a square wave—changes (is "dispersed") as its transitions are slowed down by the distributed resistance and capacitance of interconnects. Thus, the exact time difference between the analog signal and the clock edge varies from one side of the chip to the other, causing harmonic distortion in the sampled waveform.

6.3.8 Gray Encoding

Two of the potential errors in flash converters, namely, metastability and sparkles, can be suppressed using Gray encoding as an intermediate step between thermometer and binary codes. The probability of metastable states can be lowered because in Gray encoding no signal is applied to more than one input, allowing the use of pipelining to increase the time for regeneration. The effect of sparkles is reduced because the accuracy of the Gray code degrades very gradually as more sparkles appear in the thermometer code.

We illustrate these points with the aid of a 3-bit example, depicted in Figure 6.17. From the correspondence shown here we note that the Gray code output $G_3G_2G_1$ can be expressed in terms of the thermometer code as follows:

$$G_1 = T_1\overline{T_3} + T_5\overline{T_7} \tag{6.16}$$

$$G_2 = T_2\overline{T_6} \tag{6.17}$$

$$G_3 = T_4. \tag{6.18}$$

The key point in these equations is that each T_j appears in only one expression and hence no signals in the logic split. As a result, metastable errors can be reduced by pipelining the encoding, for example, as shown in Figure 6.18. Note that, for resolutions of 4 bits and above, the number of latches decreases by roughly a factor of 2 after every level of logic.

To see the robustness of Gray encoding with respect to sparkles, consider the cases illustrated in Figure 6.19. We note that while the number of sparkles increases, the Gray output remains fairly close to the top of the thermometer code, providing a reasonable approximation of the sampled value.

The performance of flash converters is determined primarily by that of their constituent comparators. Thus, the realizability of this architecture in a

Thermometer							Gray			Binary		
T_1	T_2	T_3	T_4	T_5	T_6	T_7	G_3	G_2	G_1	A	B	C
0	0	0	0	0	0	0	0	0	0	0	0	0
1	0	0	0	0	0	0	0	0	1	0	0	1
1	1	0	0	0	0	0	0	1	1	0	1	0
1	1	1	0	0	0	0	0	1	0	0	1	1
1	1	1	1	0	0	0	1	1	0	1	0	0
1	1	1	1	1	0	0	1	1	1	1	0	1
1	1	1	1	1	1	0	1	0	1	1	1	0
1	1	1	1	1	1	1	1	0	0	1	1	1

Fig. 6.17   Correspondence among thermometer, Gray, and binary codes.

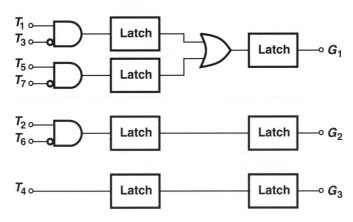

Fig. 6.18   Gray encoding with pipelining.

	Thermometer Code	Gray Code	Equivalent Decimal Output
No Sparkle:	111111111111100	1011	13
One Sparkle:	111111111111010	1000	15
Two Sparkles:	111111111111001	1010	12

Fig. 6.19   Gray encoding in the presence of sparkles.

given technology depends on the speed and accuracy with which comparisons can be performed. As a result, the high speed and superior matching of bipolar transistors have made them the dominant technology for flash ADCs. CMOS devices, with their low transconductance and large mismatch, have not yet provided a competing performance. Nonetheless, high-speed CMOS ADCs are still in demand because they can be integrated in a CMOS signal processing environment [6].

6.4 TWO-STEP ARCHITECTURES

The exponential growth of power, area, and input capacitance of flash converters as a function of resolution makes them impractical for resolutions above 8 bits, calling for other topologies that provide a more relaxed trade-off among these parameters. Two-step architectures trade speed for power, area, and input capacitance.

In a two-step ADC, first a coarse analog estimate of the input is obtained to yield a small voltage range around the input level. Subsequently, the input level is determined with higher precision within this range. Figure 6.20 illustrates a two-step architecture consisting of a front-end SHA, a coarse flash ADC stage, a DAC, a subtractor, and a fine flash ADC stage. We describe its operation using the timing diagram shown in the same figure and the waveforms of Figure 6.21.

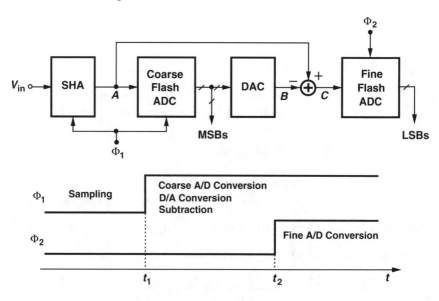

Fig. 6.20   Two-step ADC architecture.

For $t < t_1$, the SHA tracks the analog input. At $t = t_1$, the SHA enters the hold mode and the first flash stage is strobed to perform the coarse conversion. The first stage then provides a digital estimate of the signal held by the SHA (V_A), and the DAC converts this estimate to an analog signal (V_B), which is a coarse approximation of the SHA output. Next, the subtractor generates an output equal to the difference between V_A and V_B (V_C, called the "residue"), which is subsequently digitized by the fine ADC.

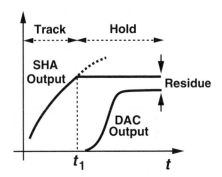

Fig. 6.21 Internal waveforms of the ADC of Figure 6.20.

If each of the two flash stages resolves $m/2$ bits, the digital output is an m-bit representation of the analog input. In practice, other combinations such as $(m/2 + 1, m/2 - 1)$ may be used for the two stages. Also, as discussed in Chapter 8, if digital correction is employed, some redundancy is added to one of the flash stages; i.e., the sum of their resolutions is slightly greater than m.

The front-end SHA plays a crucial role in the performance of two-step ADCs. Without the SHA, the maximum allowable slew rate of the input signal is severely limited. This occurs because if the analog input varies rapidly in the conversion mode, then the signal level digitized by the first stage is not equal to that sensed by the subtractor immediately before fine conversion. Figure 6.22 illustrates this timing issue, which fundamentally arises from the nonzero delay of the first stage and the DAC. For error-free conversion, $\Delta V < 1$ LSB. To quantify the resulting limitation, we assume a full-scale input $V_{in} = A \sin 2\pi f t$. If the analog estimate, V_B, settles to its proper value T_d seconds after the first stage is strobed (i.e., $T_d = t_2 - t_1$), then the analog input must vary by less than 1 LSB during T_d. Since the maximum slew rate of the input is equal to $2\pi f A$, we have

$$2\pi f A T_d < 1 \text{ LSB} \qquad (6.19)$$

$$< \frac{2A}{2^m}. \qquad (6.20)$$

In a typical case, T_d is roughly one-fourth of the total conversion period, $T_d \approx 1/(4f_S)$, where f_S is the conversion rate, and hence

$$f < \frac{f_S}{\pi 2^{m-2}}, \qquad (6.21)$$

The limitation imposed by (6.21) is substantially tighter than its theoretical counterpart in a Nyquist converter ($f < f_S/2$). For this reason, two-step architectures usually require a front-end sample-and-hold amplifier.

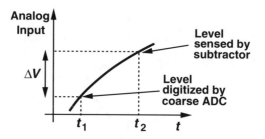

Fig. 6.22   Timing issue in the absence of front-end SHA in a two-step ADC.

The above discussion also reveals another timing issue *in the presence of* SHAs: since the output of a typical SHA takes a finite time to settle after the transition from the sampling to the hold mode, the coarse conversion cannot begin immediately after that transition if the subtractor is to sense the same level. Figure 6.23 repeats the timing diagram of Figure 6.21 with more realistic waveforms. We note that if the coarse conversion occurs at $t = t_1$, then the level digitized by the coarse stage substantially deviates from that sensed by the subtractor before the fine conversion. A simple way of avoiding this error is to begin the coarse conversion only after the SHA output has settled to within 0.5 LSB of its final value. Of course, during this time the ADC is idle, i.e., the conversion time increases by the SHA hold settling time. A better approach is to strobe the first stage before complete settling and correct the error digitally, a subject discussed in Chapter 8.

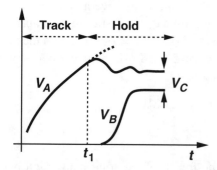

Fig. 6.23   Realistic internal waveforms of the ADC in Figure 6.21.

In addition to this problem, the front-end SHA raises other concerns that do not exist in full-flash converters. The linearity and dynamic range of the SHA directly affect those of the overall system, while the speed-precision trade-offs described in Chapters 2 and 3 limit the conversion rate. Further-more, the input capacitance and kickback noise of the coarse stage compara-tors degrade the SHA output settling behavior.

It is instructive to compare the sequence of operations in full-flash and two-step architectures (Figure 6.24). In the former type, the comparators track the input signal for approximately half of the clock cycle and perform the conversion in the other half. In the latter, while the SHA tracks the input, the rest of the system is idle (with the exception of circuits that perform offset cancellation or linearity calibration during this time). After the SHA goes into the hold mode, coarse A/D conversion, interstage D/A conversion and subtraction, and fine A/D conversion must be carried out. This comparison suggests a number of speed limitations in two-step architectures that do not exist in flash converters. First, the front-end SHA is typically much slower than a comparator, i.e., requires longer tracking and hold periods. Second, several operations must be performed in the conversion period, each of which entails speed-precision trade-offs. Even though the coarse conversion can be as fast as a flash ADC, the DAC output must settle sufficiently close to its final value, and the subtractor and fine converter require additional time.

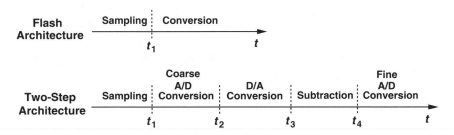

Fig. 6.24   Comparison of timing in flash and two-step architectures.

Another issue relates to the interface between the subtractor and the fine stage. If this interface does not entail any amplification, the input to the fine stage is equal to the difference between the SHA output and the DAC output, thus demanding a resolution better than 1 LSB in the fine stage. Since two-step architectures are usually used for resolutions of 10 bits and above, this implies that the comparators in the fine stage must correctly re-solve small voltages. On the other hand, if the subtractor is followed by an amplifier of gain A, the resolution required of the fine stage is relaxed by the same factor. While easing the design of comparators, this amplifier now adds a finite delay to the conversion period and contributes nonlinear-ity. More importantly, since the fine ADC must compare the amplifier out-put against a set of reference voltages, the gain A must be well-controlled so that the full-scale output of the subtractor matches the full-scale reference voltage of the second stage. This issue is discussed in more detail later.

The presence of the DAC and the subtractor in the critical signal path implies that their linearity must be commensurate with that of the overall system, while their speed must be as high as possible. These requirements have motivated the invention of various circuit techniques to improve the performance of these functions in two-step architectures [7, 8, 29].

While the front-end SHA in a two-step ADC suppresses many of the timing inaccuracies often present in converters that have no SHA, it still cannot avoid one type of error: metastability. If the SHA output voltage is very close to the reference voltage of one of the coarse stage comparators, then that comparator will have an indeterminate logical output for a long time. Depending on the type of subsequent decoding and the interstage DAC, this error may severely corrupt the analog estimate produced by the DAC, hence introducing large errors in the overall digital output. As an example, consider a section of a 10-bit two-step ADC as depicted in Figure 6.25, where the coarse stage directly drives a segmented current-steering DAC ($I_{EE}R_L = 32$ LSB). If comparator A_j is metastable, then its differential output is small and the differential pair Q_1-Q_2 steers only half of I_{EE} to the DAC output node, X. As a result, V_X deviates from its ideal value (V_j) by approximately 16 LSB. The subtractor generates the difference between V_{in} and V_X, and the fine stage digitizes this difference. Thus, an error of 16 LSB appears in the fine stage output and hence in the overall ADC output (assuming the first stage produces a correct digital output).

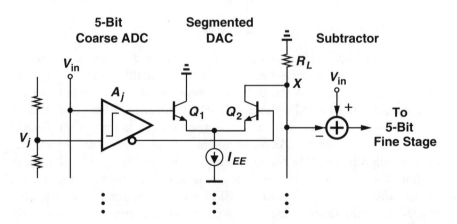

Fig. 6.25   Section of a 10-bit two-step ADC.

6.4.1 Effect of Nonidealities

We now study the effect of various nonidealities on the input/output characteristics of two-step converters. The following analysis provides more insight into the behavior of such converters and can also be used for other multistep architectures. Without loss of generality, we assume that each stage resolves 5 bits.

Consider the ADC of Figure 6.20. The input/output transfer characteristic of the coarse stage is shown in Figure 6.26, where the vertical axis represents the analog equivalent of the digital output. In the ideal case, all the code transitions occur at integer multiples of 32 LSB and the output values follow a straight line through the origin with a unity slope. In reality, the characteristic exhibits gain error (a slope different from unity), DNL (code transitions at values other than integer multiples of 32 LSB), INL (output values that do not follow a straight line), and offset (vertical shift).

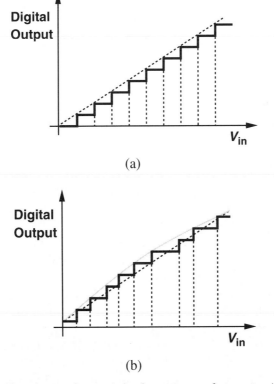

(a)

(b)

Fig. 6.26 Input/output characteristic of coarse stage of a two-step ADC. (a) Ideal; (b) actual.

In order to obtain a better view of these effects, the concept of the quantization error plot (Section 6.1) can be generalized to a residue plot wherein the difference between the actual and ideal characteristics is depicted as a function of the input (Figure 6.27). In a residue plot, gain error appears as peaks that follow a nonhorizontal straight line, DNL as shifted transition points, INL as peaks that do not follow a straight line, and offset as a vertical shift [Figure 6.27(b)].

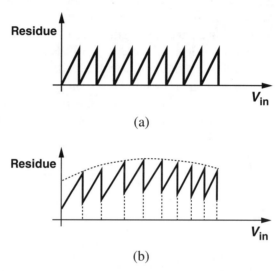

(a)

(b)

Fig. 6.27   Residue plot derived from characteristic of Figure 6.26. (a) Ideal; (b) including errors.

In a two-step ADC, the residue characteristic can be studied at three different interfaces: the output of the coarse stage, the output of the DAC, and the output of the subtractor. The nonidealities introduced by each of these stages influence the residue, hence demanding a careful allocation of errors so as to maintain the overall DNL and INL below 0.5 LSB.

If the coarse flash stage employs a resistor ladder to subdivide the reference, gain error in that stage is zero because the end points of the transfer characteristic are defined by the voltages applied to the end points of the ladder. However, DNL, INL, and offset still exist due to the offset of comparators and mismatch of the ladder's resistors.

The interstage DAC may introduce DNL, INL, gain error, and offset in the residue. Resistor ladder and current-steering DACs exhibit DNL, INL, and gain error (if their full-scale range is not exactly equal to that of the second stage). In capacitor DACs, all four errors may exist because precharge switches induce offset at the output as well (Chapter 4).

The interstage subtractor introduces offset and gain error in the residue signal. The latter is particularly important because it leads to DNL in the overall converter. Consider the residue plot of Figure 6.28, where the output of a typical subtractor with a gain slightly less than unity is shown. If V_{in} is slightly *less* than V_j, then the subtractor produces an output equal to 30 LSB (rather than 32 LSB). The second stage digitizes this difference, and the result is added to the output of the first stage, V_{j-1}. On the other hand, if V_{in} is slightly *greater* than V_j, the subtractor output is nearly zero and the overall output is equal to V_j $(= V_{j-1} + 32\ \text{LSB})$. Thus, the overall input/output characteristic exhibits an unwanted jump at $V_{in} = V_j$ [Figure 6.28(b)], making it impossible for the ADC to generate a digital output corresponding to $V_{j-1} + 31\ \text{LSB}$. This is called a "missing code."

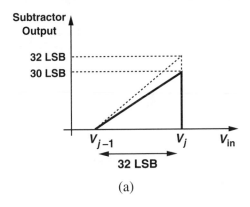

(a)

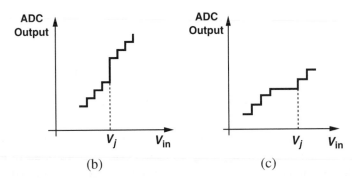

(b)   (c)

Fig. 6.28   (a) Subtractor output in a two-step ADC; (b) missing code; (c) 1-LSB DNL.

It can be easily shown that if the subtractor gain is *greater* than its ideal value, then the ADC exhibits nonmonotonicity rather than missing codes [Figure 6.28(c)].

The above discussion shows that missing codes and nonmonotonicity are avoided if the subtractor full-scale output matches the second-stage full-scale input, i.e., if the second-stage full-scale reference experiences the same gain error as does the subtractor. This can be accomplished by passing the second-stage full-scale reference through a circuit replicated from the subtractor [9, 29].

6.4.2 Two-Step Recycling Architecture

A two-step ADC need not employ two separate flash stages to perform the coarse and fine conversions; one stage can be used for both. Called the "recycling architecture" [17], such a topology is illustrated in Figure 6.29. Here, during the coarse conversion the flash stage senses the front-end SHA output, V_A, and generates the coarse digital output. This output is then converted to analog by the DAC and subtracted from V_A by the subtractor. During fine conversion, the subtractor output is digitized by the flash stage. Note that in this phase, the ADC full-scale voltage must be equal to that of the subtractor output. Thus, for proper fine conversion, either the ADC reference voltage must be reduced or the residue must be amplified.

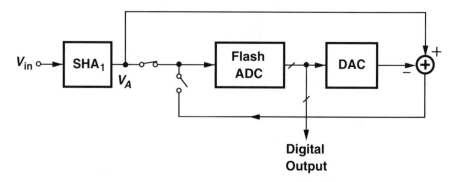

Fig. 6.29   Recycling ADC architecture.

While reducing area and power dissipation by roughly a factor of 2 relative to two-stage ADCs, recycling converters suffer from other limitations. The converter must now employ either low-offset comparators (if the subtractor has a gain of 1), inevitably slowing down the coarse conversion, or a high-gain subtractor, thereby increasing the interstage delay. This is in contrast with two-stage ADCs, where the coarse stage comparators need not have a high resolution and hence can operate faster.

6.4.3 Two-Step Subranging Architecture

A variant of two-step architectures that does not require an explicit subtractor is the subranging topology [10]. In this architecture, the coarse stage identifies and subdivides a reference voltage range around the input voltage, and the fine stage compares the input against this new set of references. The subranging technique can be explained using the characteristics shown in Figure 6.30. Again, we assume a 10-bit converter having two 5-bit flash stages. Plotted in Figure 6.30(a) is the output of the interstage DAC in an ideal two-step ADC as a function of the analog input. The estimate generated by the DAC must be subtracted from the input to produce a residue for fine quantization. Now, suppose the DAC can provide *two* different outputs, one shifted up by 32 LSB with respect to the other, as illustrated in Figure 6.30(b). Then, the input signal is always greater than V_{D1} and less than V_{D2}, and if the difference between V_{D1} and V_{D2} is subdivided by 32, a set of 32 reference voltages around the input signal will be provided. The fine stage can therefore compare the input against these reference voltages and determine the least significant bits.

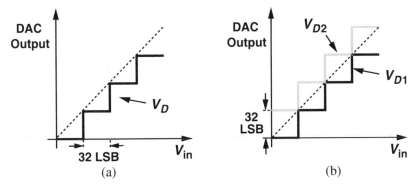

Fig. 6.30   (a) Output of interstage DAC in an ideal two-step ADC; (b) DAC outputs in a subranging architecture.

The subranging technique can be easily implemented using resistor-ladder DACs [10, 11]. Shown in Figure 6.31 is an example where two resistor ladders perform the subranging function. The first ladder divides the main reference by 32, providing segments of 32 LSB between consecutive taps. The 1-of-n code generated by the coarse stage turns on two switches connected to two consecutive taps of this ladder, thereby establishing their voltages at nodes X and Y. Buffers B_1 and B_2 provide a low driving impedance for the second ladder, which subdivides $V_{D2} - V_{D1}$ by another factor of 32 and hence produces 32 references around V_{in}.

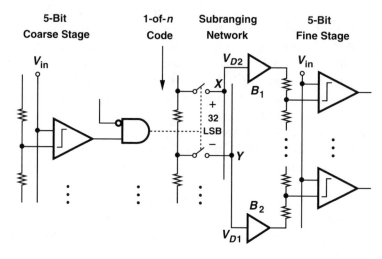

Fig. 6.31 Section of a 10-bit subranging ADC architecture.

While avoiding the difficulties associated with subtractor design, the subranging architecture suffers from two drawbacks. First, its interstage processing is slow because, as discussed in Chapter 4, resistor-ladder DACs generally exhibit long settling times. Second, the fine stage comparators must operate across the full common-mode range of the input signal while maintaining a constant, small input offset. In two-step architectures using subtractors, on the other hand, the input common-mode level of the second stage is fixed by the subtractor.

In a typical two-step ADC, it is possible to use the reference ladder of the coarse stage as the interstage D/A converter as well [18]. While saving area and power dissipation, this approach may cause the kickback noise generated by the coarse stage comparators to corrupt the sensitive output of the DAC, thus mandating longer settling times.

6.5 INTERPOLATIVE AND FOLDING ARCHITECTURES

The large input capacitance, high power dissipation, stringent timing requirements, and large area of full-flash architectures prohibit their use in many applications. A number of circuit techniques have been proposed to alleviate these problems while maintaining the "one-step" nature of the architecture, i.e., without adding sample-and-hold circuits to the ADC. Among these techniques, interpolation and folding have proved quite beneficial. While applied predominantly to bipolar circuits, these concepts have recently been used in CMOS technology as well [19, 20].

6.5.1 Interpolation

In order to reduce the number of preamplifiers at the input of a flash ADC, the *difference* between the analog input and each reference voltage can be quantized at the output of each preamplifier. This is possible because of the finite gain—and hence nonzero linear input range—of typical preamplifiers used as the front end of comparators.

We illustrate this concept by means of Figure 6.32. In Figure 6.32(a), preamplifiers A_1 and A_2 compare the analog input with V_{r1} and V_{r2}, respectively. In Figure 6.32(b), the input/output characteristics of A_1 and A_2 are shown. Assuming zero offset for both preamplifiers, we note that $V_{X1} = V_{Y1}$ if $V_{in} = V_{r1}$, and $V_{X2} = V_{Y2}$ if $V_{in} = V_{r2}$. More importantly, $V_{X2} = V_{Y1}$ if $V_{in} = V_m = (V_{r1} + V_{r2})/2$; i.e., the *polarity* of the difference between V_{X2} and V_{Y1} is the same as that of the difference between V_{in} and V_m.

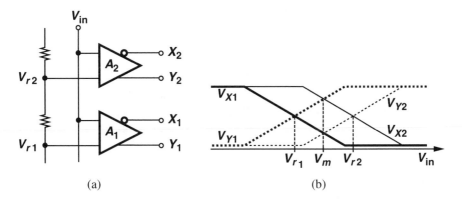

Fig. 6.32   Interpolation between outputs of two amplifiers.

The above observation indicates that the equivalent resolution of a flash stage can be increased by "interpolating" between the outputs of preamplifiers. For example, Figure 6.33 shows how an additional latch detects the polarity of the difference between single-ended outputs of two adjacent preamplifiers [21]. Note that in contrast with a simple flash stage, this approach halves the number of preamplifiers but maintains the same number of latches.

The interpolation technique of Figure 6.33 substantially reduces the input capacitance, power dissipation, and area of flash converters, while pre-

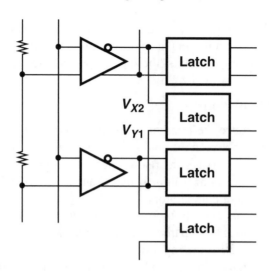

Fig. 6.33   Interpolation in a flash ADC.

serving the one-step nature of the architecture. This is possible because all of the signals arrive at the input of the latches simultaneously and hence can be captured on one clock edge. Since this configuration doubles the effective resolution, we say it has an interpolation factor of 2.

    Another interesting property that accompanies this interpolation technique is the improved differential nonlinearity due to distribution of errors [21, 23]. Figure 6.34 illustrates this point by comparing the effect of a 1-LSB offset in full-flash and in interpolative architectures. While in the first case [Figure 6.34(a)] the offset gives rise to a DNL = 1 LSB (as well as a missing code), in the second case [Figure 6.34(b)] it yields a maximum DNL of only 0.5 LSB.

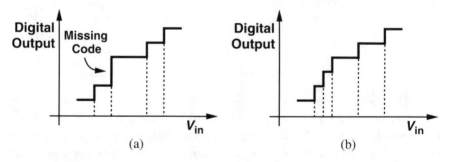

Fig. 6.34   Differential nonlinearity in (a) flash and (b) interpolative converters.

    The above interpolation approach is possible only because the front-end preamplifiers of comparators have a finite gain. To investigate this point further, let us consider the difference between V_{X2} and V_{Y1} in Figure 6.32. Plotted in Figure 6.35 is this difference for several values of $V_{r2} - V_{r1}$. As $V_{r2} - V_{r1}$ exceeds the range across which each preamplifier exhibits a nonzero gain, a "dead band" appears around $V_{in} = (V_{r1} + V_{r2})/2$ where the gain is quite small and $V_{Y1} \approx V_{X2}$. If the analog input level falls in this band, then $V_{X2} - V_{Y1}$ may not be sufficient to overcome the offset of the following latch, thereby yielding incorrect polarity for the difference between V_{in} and $(V_{r1} + V_{r2})/2$.

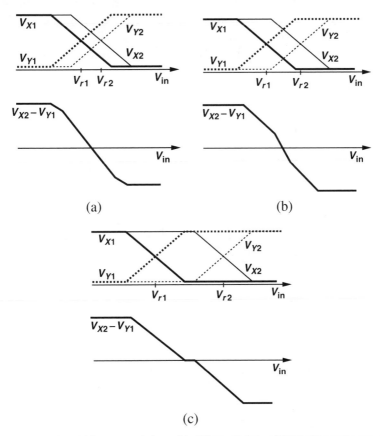

Fig. 6.35   Interpolation with different values of $V_{r2} - V_{r1}$.

In reality, the transition of input/output characteristics from high gain to low gain is not as abrupt as indicated in Figure 6.35. If the preamplifiers are implemented as bipolar differential pairs, then the dead band begins to appear when $V_{r2} - V_{r1}$ exceeds approximately $5kT/q$. If a wider range is required, each differential pair can incorporate emitter degeneration.

The concept of interpolation can be extended so as to produce more quantization levels between every two consecutive reference voltages in a flash converter, further reducing the number of input preamplifiers. For example, consider the circuit of Figure 6.36(a), where the outputs of two preamplifiers are interpolated using two uniform resistor strings [22, 23]. As illustrated in Figure 6.36(b), since the input/output characteristics of the two preamplifiers are offset by $V_{r2} - V_{r1}$, as V_{in} goes from below V_{r1} to above V_{r2}, the differential output voltages $V_{O1}, \ldots, V_{O5}$ cross zero at $V_{in} = V_{r1} + k(V_{r2} - V_{r1})/4$, for $k = 0, \ldots, 4$, respectively. Thus, if latches are used to detect the polarity of $V_{O1}, \ldots, V_{O5}$, this configuration provides an interpolation factor of 4. Note that the number of latches is still the same as in a full-flash architecture and $V_{r2} - V_{r1}$ has the same upper bound (to avoid dead bands) as described above.

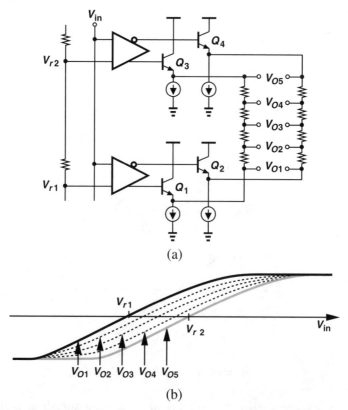

Fig. 6.36   Higher order interpolation. (a) Implementation; (b) input/output characteristics.

The technique of Figure 6.36(a) must deal with several design issues that limit the interpolation factor. First, the resistor strings and the input capacitance of the following latches introduce a time constant in the signal path, thereby reducing the bandwidth. This reduction is proportional to the *square* of the interpolation factor and hence becomes substantial if this factor exceeds ~ 4. Second, in the circuit of Figure 6.36(a), as V_{O1} and V_{O5} vary, the bias current of emitter followers that drive the resistor strings also varies, thus changing their base-emitter voltage and causing nonuniformity between zero-crossing points of the output voltages. This nonuniformity is equivalent to differential nonlinearity when referred to the input. Note that the first problem can be alleviated by reducing the value of interpolation resistors, but at the cost of exacerbating the second problem or increasing the power dissipation.

The nonlinearity error described above decreases markedly if a large number of amplifiers are used in the interpolation [23]. As illustrated in Figure 6.37, most of the current flowing through the resistor strings is provided by the amplifiers whose reference voltage is far from the input voltage. Thus, the amplifiers whose reference voltage is close to the input level need not provide the resistor string current. In order to create this effect at the two ends of the interpolation array, a few dummy preamplifiers and interpolation resistors can be added to both ends [23].

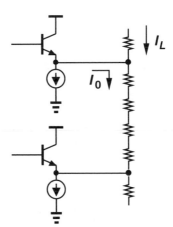

Fig. 6.37   Resistor string current in an interpolative ADC.

Interpolation techniques can also be applied to the design of CMOS ADCs. Since, as explained in Chapter 7, simple CMOS differential pairs suffer from large offset and small gain, the interpolation scheme is better implemented using autozeroed amplifiers and capacitors.

Illustrated in Figure 6.38 is an interpolative architecture employing autozeroed CMOS inverters as amplifiers [20]. The circuit has two modes of operation. In the sampling/reset mode, feedback switches S_5 and S_6 are on, driving the inverters into their high-gain region, and sampling switches S_1 and

S_3 are also on, allowing the analog input to be sampled on C_1 and C_2. At the end of the sampling mode, S_1, S_3, S_5, and S_6 turn off and S_2 and S_4 turn on, thereby changing the voltage at nodes X_1 and X_2 by $V_{in} - V_{r1}$ and $V_{in} - V_{r2}$, respectively. This voltage change is amplified by each inverter and combined with adjacent ones by the interpolation capacitors, yielding an interpolation factor of 2. In practice, the sampling and evaluation modes can be pipelined so as to increase the conversion rate [20].

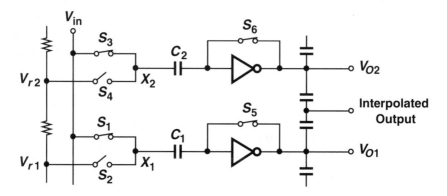

Fig. 6.38   Interpolation in a CMOS ADC.

6.5.2 Folding Architectures

Folding architectures have evolved from flash and two-step topologies. As explained in Section 6.3, flash architectures provide a one-step operation with no need for analog postprocessing, but they suffer from large input capacitance, large power dissipation, and severe timing problems as speed and resolution increase. Two-step architectures, on the other hand, have much less hardware but require a front-end sample-and-hold circuit as well as analog postprocessing. Folding architectures perform analog *preprocessing* to reduce the hardware while maintaining the one-step nature of flash architectures.

The basic principle in folding is to generate a residue voltage through analog preprocessing and subsequently digitize that residue to obtain the least significant bits. The most significant bits can be resolved using a coarse flash stage that operates *in parallel* with the folding circuit and hence samples the signal at approximately the same time that the residue is sampled. Figure 6.39 depicts the generation of residue in two-step and folding architectures. In a two-step architecture, coarse A/D conversion, interstage D/A conversion, and subtraction must be completed before the proper residue becomes available.

In contrast, folding architectures generate the residue "on the fly" using simple wideband stages.

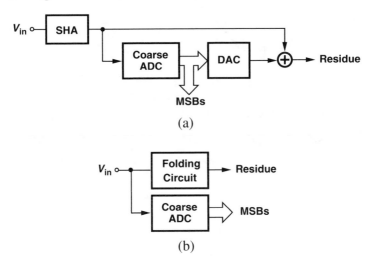

Fig. 6.39   Generation of residue in (a) two-step and (b) folding architectures.

To illustrate the above principle, we first describe a simple, ideal approach to folding. Consider two amplifiers A_1 and A_2 with the input/output characteristics depicted in Figure 6.40(a). The active region of one amplifier is centered around $(V_{r2} + V_{r1})/2$ and that of the other around $(V_{r3} + V_{r2})/2$, and $V_{r3} - V_{r2} = V_{r2} - V_{r1}$. Each amplifier has a gain of 1 in the active region and 0 in the saturation region. If the outputs of the two amplifiers are summed, the "folding" characteristic of Figure 6.40(b) results, yielding an output equal to $V_{in} - V_{r1}$ for $V_{r1} < V_{in} < V_{r2}$ and $-V_{in} + V_{r2} + \Delta$ for $V_{r2} < V_{in} < V_{r3}$, where Δ is the value of the summed characteristics at $V_{in} = V_{r2}$. If V_{r1}, V_{r2}, and V_{r3} are the reference voltages in an ADC, then these two regions can be viewed as the residue characteristics of the ADC for $V_{r1} < V_{in} < V_{r3}$. To understand why, we compare this characteristic with that of a two-step architecture, as shown in Figure 6.40(c). The two characteristics are similar except for a negative sign and a vertical shift in the folding output for $V_{r2} < V_{in} < V_{r3}$. Thus, if the system accounts for the sign reversal and level shift, the folding output can be used as the residue for fine digitization.

Shown in Figure 6.41 is an implementation of folding. Here, four differential pairs process the difference between V_{in} and $V_{r1}, \ldots, V_{r4}$, and their output currents are summed at nodes X and Y. Note that the outputs of adjacent stages are added with opposite polarity; e.g., as V_{in} increases, Q_1 pulls node X low while Q_2 pulls node Y low. Current source I_5 shifts V_Y down by IR.

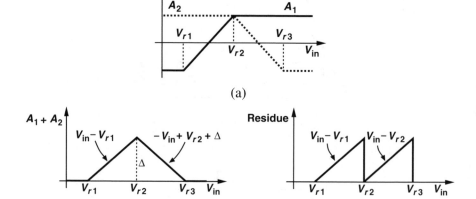

Fig. 6.40   (a) Input/output characteristics of two amplifiers; (b) sum of characteristics in (a); (c) residue in a two-step ADC.

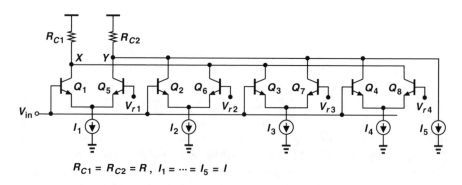

Fig. 6.41   Folding circuit.

To explain the operation of the circuit, we consider its input/output characteristic, plotted in Figure 6.42. For V_{in} well below V_{r1}, Q_1-Q_4 are off, Q_5-Q_8 are on, I_2 and I_4 flow through R_{C1}, and I_1, I_3, and I_5 flow through R_{C2}. As V_{in} increases, Q_1 begins to turn on, while Q_2-Q_4 remain off (if $V_{r1}, \ldots, V_{r4}$ are sufficiently far from each other). For $V_{in} = V_{r1}$, Q_1 and Q_5 share I_1 equally, yielding $V_X = V_Y$. As V_{in} exceeds V_{r1} by several V_T, Q_5 turns off, allowing V_X and V_Y to reach V_{min} and V_{max}, respectively. As V_{in} approaches V_{r2}, Q_2 begins to turn on and the circuit behaves in a similar manner as before. Considering the differential output, $V_X - V_Y$, we note that the resulting characteristic exhibits folding points at $(V_{r1} + V_{r2})/2$, $(V_{r2} + V_{r3})/2$, etc. As V_{in} goes from below V_{r1} to above V_{r4}, the slope of $V_X - V_Y$ changes sign four times; hence we say the circuit has a folding factor of 4.

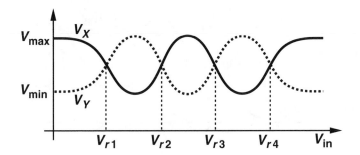

Fig. 6.42   Folding characteristic of the circuit in Figure 6.41.

The simplicity and speed of folding circuits have made them quite popular in A/D converters, particularly because they eliminate the need for sample-and-hold amplifiers, D/A converters, and subtractors. Nevertheless, these circuits suffer from several drawbacks that limit their use, especially for resolutions above 6 bits. We describe a number of these drawbacks here.

In the folding characteristic of Figure 6.42, if the input goes from zero to full scale once, the output goes from V_{min} to V_{max} four times; i.e., a folding factor of n results in a frequency multiplication by n. Thus, the bandwidth required of the folding circuit is n times that of the maximum input frequency, thereby imposing strong trade-offs among speed, gain, and power dissipation. As a consequence, in high-speed systems the folding factor is typically between 2 and 4 [24].

Another property of the folding characteristic shown in Figure 6.42 is its substantial nonlinearity. This can be better seen by comparing this characteristic with an ideal folding relation as depicted in Figure 6.43. The deviation of the actual characteristic from a straight line translates into differential nonlinearity and is plotted in the same figure. For a typical design, the maximum deviation can be as high as several tens of millivolts, prohibiting the use of simple folding even for resolutions as low as 8 bits.

The difference between V_{rj} and $V_{r(j+1)}$ in Figure 6.41 has significant impact on the input/output characteristics. In analyzing the folding circuit of Figure 6.41, we assumed that as V_{in} approaches V_{rj}, only the differential pair having that reference switches while others retain the same state. This holds only if $V_{r(j+1)} - V_{rj}$ is sufficiently large. On the other hand, as with interpolation, this difference cannot be arbitrarily large because a dead band with low gain appears in the characteristic. The optimum difference is approximately equal to $5V_T$ [22].

The nonlinearity errors in folding characteristics also depend on the frequency of operation. At high speeds, the rate of change of signals becomes comparable with the intrinsic time constants of the circuit, thus causing

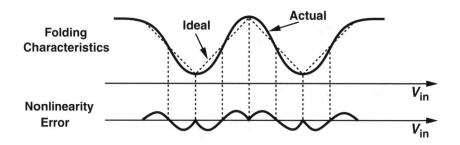

Fig. 6.43   Nonlinearity resulting from folding.

"rounding" of the characteristic at the folding points and hence increasing the nonlinearity.

Even though some converters solely based on the above folding technique have been designed [26, 27, 28], other methods have been devised so as to exploit the features of folding without incurring excessive nonlinearity. In particular, folding and interpolation can be combined to achieve efficient, high-speed A/D conversion.

6.5.3 Folding with Interpolation

Before describing architectures that combine folding and interpolation, we should make two important observations. First, in the folding characteristic of Figure 6.42, the nonlinearity falls to zero at zero-crossing points. Thus, if only these points are considered, the *polarity* of the difference between V_{in} and V_{rj} can be determined correctly. To resolve the lower bits, additional zero-crossing points can be produced using interpolation [22]. Second, if the folding circuit of Figure 6.41 is replicated and its reference voltages are shifted by $(V_{r(j+1)} - V_{rj})/2$ [Figure 6.44(a)], then a second folding characteristic is obtained as shown in Figure 6.44(b) [28]. We call the original characteristic the "in-phase" (I) output and the second the "quadrature" (Q) output [29]. Note that in a fully differential implementation, the inverted versions of these outputs ($\overline{I}$ and $\overline{Q}$, respectively) are also available. For simplicity, we call each of these characteristics a "folding signal."

We now describe how I and Q outputs of a double-folding circuit can be used in an interpolation network to generate additional zero-crossing points corresponding to lower bits. Consider the circuit of Figure 6.45, where the I and Q outputs of a folding circuit are applied to four emitter followers and a resistor network, providing an interpolation factor of 2 by generating additional differential voltages V_{ac} and V_{bd}. Note that the symmetry of the interpolation network suppresses variation of V_{BE} of the emitter followers as the input varies [29].

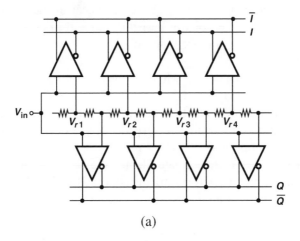

(a)

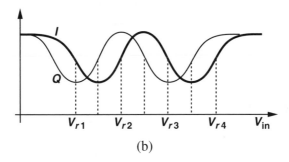

(b)

Fig. 6.44   Folding with in-phase and quadrature outputs.

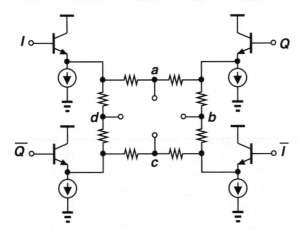

Fig. 6.45   Interpolation circuit for in-phase and quadrature signals.

The interpolation factor of the above circuit can be increased by increasing the number of the interpolation resistors, but at the cost of increasing the output time constant. In practice, interpolation factors as high as 8 have been utilized [29].

It is interesting to note that while the idea of folding was originally conceived to produce a residue signal whose *amplitude* could be finely digitized (as in a two-step architecture), the combination of folding and interpolation extracts information only from *zero-crossing points* of the residue with little concern about its amplitude.

For interpolation factors greater than 2, the deviation of folding signals from an ideal triangular waveform still introduces differential nonlinearity. As shown in Figure 6.46 for an interpolation factor of 4, while the zero-crossing points of $V_{2/4}$ coincide with their ideal values, those of $V_{1/4}$ and $V_{3/4}$ deviate from their ideal values because the top (or bottom) portion of a folding signal (where deviation from an ideal ramp is maximum) is linearly combined with the middle portion of another folding signal (where deviation is minimum).

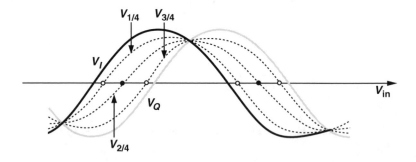

Fig. 6.46 Deviation of interpolated zero crossings due to folding nonlinearity.

The above observation may suggest that the DNL can be reduced if the folding signals are brought closer together (i.e., the difference between V_{rj} and $V_{r(j+1)}$ is reduced) because the interpolation is then performed on the linear portions of the folding signals. However, this violates one of the assumptions originally made in deriving the folding signals of Figure 6.42: when one differential pair begins to turn on, its adjacent stages are not completely off and contribute to the output variation [22]. Thus, the difference between V_{rj} and $V_{r(j+1)}$ must be chosen so as to minimize the overall DNL. Simulations indicate that this error reaches a minimum for $V_{r(j+1)} - V_{rj} \approx 5V_T$ [22]. This error can also be reduced by *nonlinear* interpolation [29], i.e., using unequal resistors in the interpolation network.

Several different implementations of folding and interpolation architectures have been reported [22, 24, 29]. In addition to the circuit of Figure 6.41, the topology shown in Figure 6.47 is frequently used. This circuit consists of four differential pairs whose outputs are combined using four emitter followers. The two sets of interconnected emitter followers in essence constitute two "analog wired-OR" circuits so that their outputs go high if the collector voltage of one of the differential pairs goes high. While in the folding circuit of Figure 6.41 several collectors are connected to the same node, thus contributing substantial capacitance at that node, the circuit of Figure 6.47 does not suffer from this drawback and drives the summing node by emitter followers.

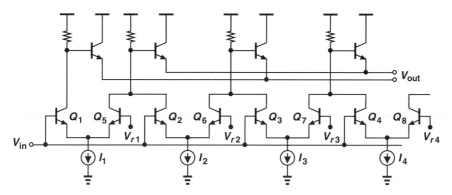

Fig. 6.47   Folding circuit with analog wired-OR.

A critical issue in folding ADCs is the timing error between the coarse stage and the folding amplifier. Since these two circuits are inherently different, they introduce unequal delays in the analog signal, thus presenting slightly different points of the input to the subsequent latches. As a result, the coarse stage may "point" to the wrong cycle in the folding characteristics. Various correction techniques are often utilized to eliminate this error [24, 25].

Folding architectures have been implemented primarily in bipolar technology. The small offset, high-speed switching, and exponential I-V characteristics of bipolar transistors allow resolutions of 8 bits in inherently one-step architectures, achieving conversion rates above 600 MHz [24]. In CMOS technology, on the other hand, large offsets, low transconductance, and short-channel effects make it difficult to achieve resolutions above approximately 6 bits without offset cancellation. For example, a CMOS interpolating ADC built in a 1-μm technology requires an 8-V supply to maintain an LSB voltage greater than the input offset of its constituent comparators [19].

6.6 PIPELINED ARCHITECTURES

The concept of pipelining, often used in digital circuits, can also be applied in the analog domain to achieve higher speed where several operations must be performed serially. Figure 6.48 shows a general (analog or digital) pipelined system. Here, each stage carries out an operation on a sample, provides the output for the following sampler, and, once that sampler has acquired the data, begins the same operation on the next sample. Thus, at any given time, all the stages are processing different samples concurrently, and hence the throughput rate depends only on the speed of each stage and the acquisition time of the next sampler. (Note the analogy between a pipelined system and an assembly line.)

Fig. 6.48   A pipelined system.

To arrive at a simple example of analog pipelining, consider a two-step ADC where four operations, namely, coarse A/D conversion, interstage D/A conversion, subtraction, and fine A/D conversion, must be performed serially. As such, the ADC cannot begin to process the next sample until all four operations are finished. Now, suppose a SHA is interposed between the subtractor and the fine stage, as shown in Figure 6.49, so that the residue is stored before fine conversion begins. Thus, the front-end SHA, the coarse ADC, the interstage DAC, and the subtractor can start processing the next sample while the fine ADC operates on the previous one, thereby allowing potentially faster conversion.

Pipelining can be employed even more extensively than described in the above example. Shown in Figure 6.50 is a more general form of pipelined ADCs. The architecture consists of N stages, each including a SHA, an ADC, a DAC, a subtractor, and possibly an amplifier, with actual implementations often combining two or more of these functions in one circuit. A typical conversion proceeds as follows. The first stage samples and holds the analog input, produces a k-bit digital estimate of the held input, converts this estimate to analog, subtracts the result from the held input, and in some implementations amplifies the residue (by a power of 2). Next, the following stage in the pipeline samples the (amplified) residue and performs the same sequence of operations while the first stage begins processing the next sample. As every stage incorporates a sample-and-hold function, the analog data is preserved,

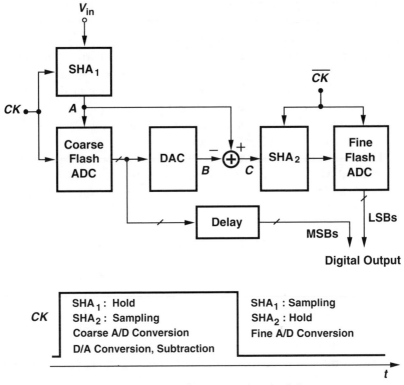

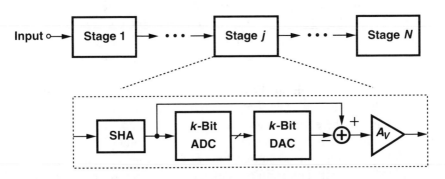

Fig. 6.49 Two-step ADC employing pipelining.

Fig. 6.50 General pipelined ADC architecture.

allowing different stages to process different samples concurrently. Thus, the conversion rate depends on the speed of only one stage, usually the front end.

While the concurrent operation of pipelined converters makes them attractive for high speeds, their extensive linear processing of the analog input relies heavily on operational amplifiers, which are relatively slow building blocks in analog design. The maximum allowable gain error and nonlinearity of the SHA and residue amplifier in each stage is commensurate with the number of bits resolved afterward and must be maintained well below 1 LSB in the first few stages, thus mandating the use of high-gain op amps in high-resolution converters. Single-supply op amp designs with such gains and large output swings suffer from slew rate and phase margin degradation.

The number of bits resolved in each stage and hence the number of stages of a pipelined ADC depend on various considerations such as overall resolution, speed, and technology. Practical implementations vary from cascaded flash stages to 1-bit-per-stage topologies [30, 31], each of which has its own merits and drawbacks.

While flash stages relax the linearity and gain requirements of the following SHAs by the number of bits they resolve, they also necessitate the use of explicit, precise D/A converters. On the other hand, pipelined architectures that resolve 1 bit in each stage lend themselves to a simple realization, as shown in Figure 6.51 [31]. Illustrated in single-ended form, the circuit corresponds to one stage of a pipelined ADC and consists of nominally equal capacitors C_1 and C_2 and op amp A. A typical conversion cycle proceeds as follows. First, the circuit is configured as in Figure 6.51(a), with the input sampled on C_1 and C_2. Next, the dc loop around A opens, C_1 is placed around A, and C_2 is switched to V_{REF}. The output voltage is then equal to $2V_{in} - V_{REF}$. Subsequently, the polarity of this output is detected to determine whether $V_{in} > V_{REF}/2$ or $V_{in} < V_{REF}/2$ and hence the corresponding bit. If $V_{in} < V_{REF}/2$, V_{REF} is added back to the residue, yielding $V_{out} = 2V_{in}$, which in the next stage is amplified by another factor of 2 and compared with V_{REF} again [31].

Rather than subdividing the reference by means of a DAC, this topology amplifies the residue by a factor of 2, thereby demanding well-matched gain-setting elements (here, capacitors) and high-gain op amps. In order to achieve a precise gain of 2, averaging techniques [31] or calibration [32] can be used so as to improve the equivalent matching of capacitors (Chapter 8).

Pipelined architectures, in principle, require less power dissipation and area than full-flash configurations. This is especially evident in the 1-bit-per-stage topology, where area and power grow *linearly*—rather than exponentially—with the number of bits. In practice, however, the op amps utilized in each stage employ large devices biased at high currents to attain a high speed with a wide dynamic range (i.e., large input and output voltage swings and low noise).

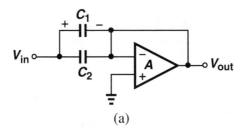

(a)

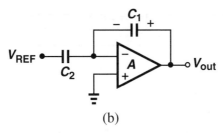

(b)

Fig. 6.51   One-bit-per-stage pipelined operation. (a) Sampling; (b) D/A conversion, subtraction, and amplification.

Pipelined architectures rely heavily on the storage and processing of analog signals. Thus, they can be easily implemented in CMOS or BiCMOS technologies, where simple sampling switches and high-input-impedance devices are readily available. This applies particularly to the 1-bit-per-stage configuration, which indeed has evolved in CMOS technology. At the 10-bit resolution level, sampling rates of 100 MHz in BiCMOS [33] and 50 MHz in CMOS [30] have been achieved.

6.7 SUCCESSIVE APPROXIMATION ARCHITECTURES

Successive approximation employs a "binary search" algorithm in a feedback loop including a 1-bit A/D converter. Figure 6.52 illustrates this architecture, which consists of a front-end SHA, a comparator, a pointer (shift register), decision logic, a decision register, and a DAC. The pointer points to the last bit changed in the decision register, and the data stored in this register is the result of all the comparisons performed in the present conversion period.

During a binary search, the circuit halves the difference between the held signal, V_H, and the DAC output, V_D, in each clock cycle. The conversion proceeds as follows. First, both the pointer and the decision register are set to midscale (100...0) so that the DAC produces the midscale analog output. The comparator is then strobed to determine the polarity of $V_H - V_D$. The pointer and decision logic direct the logical output of the comparator to the

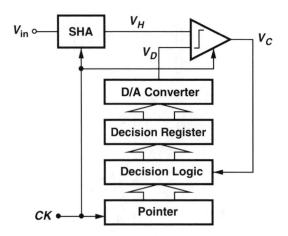

Fig. 6.52   Successive approximation architecture.

most significant bit (MSB) in the decision register. Thus, if $V_H > V_D$, the MSB of this register is maintained at ONE, and if $V_H < V_D$, it is set to ZERO. Subsequently, the pointer is set to $110\ldots 0$ and the next bit in the decision register is set to ONE. After the DAC output has settled to its new value, the comparator is strobed again and the above sequence is repeated. Figure 6.53 illustrates the DAC output waveform in a typical conversion period.

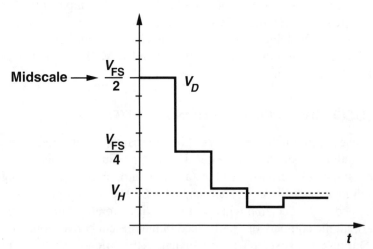

Fig. 6.53   DAC output waveform in Figure 6.52.

For a resolution of M bits, the successive approximation architecture is at least M times slower than full-flash configurations, but it offers several advantages. First, note that the comparator offset voltage does not affect the linearity of the overall converter because it can be represented as a voltage source in series with the SHA output, indicating that the offset voltage simply adds to the analog input and hence appears as an offset in the overall characteristics. Consequently, the comparator can be designed for high-speed operation even in high-resolution systems. Of course, the input rms noise of the comparator must be much less than 1 LSB. Second, this architecture does not require an explicit subtractor, an important advantage in high-resolution applications. Third, the circuit complexity and power dissipation are in general less than that of other architectures.

If the front-end SHA provides the required linearity and speed and the comparator input-referred noise is small enough, then the converter's performance depends primarily on that of the DAC. In particular, differential and integral nonlinearity of the converter are given by those of the DAC, and the maximum conversion rate is limited by its output settling time. Note that in the first conversion cycle, the DAC output must settle to maximum resolution of the system so that the comparator determines the MSB correctly. Thus, if the clock period is constant, the following conversion cycles will be as long as the first one, implying that the conversion rate is constrained by the speed of the DAC.

Successive approximation converters that incorporate capacitor DACs are usually based on the "charge redistribution" principle [35]. We illustrate this principle using the simplified diagram of Figure 6.54, where the DAC consists of binary-weighted capacitors C_1-C_M ($C_j = 2C_{j-1}, j = M, \ldots, 2$) and $C_0 = C_1$ [35]. In the sampling mode, the top plate is grounded, while all the bottom plates are connected to the input signal. In the transition from the sampling mode to the hold/conversion mode, S_P turns off and all the bottom plates are grounded, causing the top plate voltage to be equal to the negative of the sampled level. The conversion then proceeds by switching the bottom plate of some of the C_1-C_M to V_{REF} (according to a binary search algorithm) such that the top plate voltage eventually returns to zero. For example, to evaluate the most significant bit, the bottom plate of C_M is switched from ground to V_{REF} so that the top plate voltage increases by $V_{REF}/2$. Subsequently, the comparator is strobed to determine the polarity of the difference between the top plate voltage and ground, and hence the MSB. The following steps are similar to those described for Figure 6.52.

The circuit of Figure 6.54 has several interesting features. Here, the D/A converter operates as a sample-and-hold circuit as well: the capacitor array acts as the storage element, while the top and bottom plate switches control the sampling. The accuracy-speed trade-off described for MOS switches in Chapter 2 is considerably relaxed here because the sample-to-hold transition

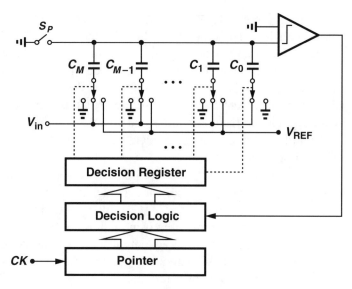

Fig. 6.54   Charge redistribution architecture.

is performed by S_P, which always turns off with its source and drain at ground potential, injecting a constant charge onto the array. This is in contrast with the sampling circuit of Figure 2.6, where the charge injected by the switch is input-dependent.

Another feature of this configuration is that at the end of the conversion, the top plate potential is very close to zero. This in turn means that the junction capacitance of S_P contributes very little nonlinearity to the system because its net voltage change is nearly zero [35]. Note that since during sampling S_P is in series with the entire array, it must have a low on-resistance—and hence large width—so as to provide a fast acquisition. Consequently, its junction capacitance is usually comparable with the value of the smallest capacitor in the array. Also, because the comparator eventually compares V_D with the ground potential, it need not maintain a high precision across a wide input common-mode range, an important feature in low-voltage designs. Nonetheless, as discussed in Chapter 7, the comparator must have a fast overdrive recovery, i.e., must not slow down after it has sensed a large *differential* input.

For high resolutions, the ratio of the largest and the smallest capacitors (2^{m-1}), as well as the total value of the array capacitance, can be excessively large. For example, in a 12-bit converter, the ratio of the MSB and to the LSB capacitors is equal to 2048, and the array comprises 4096 equal unit capacitors. As the minimum size of the smallest capacitor is often dictated by uniformity and matching considerations, the area and capacitance of such an array may be very large, thus yielding an enormous input capacitance for the converter, slowing down the preceding circuit, and complicating the routing. Additionally, the large capacitance of the array draws large current spikes from the ground and V_{REF} lines during transients, causing ringing and long settling times in the presence of inductance in series with these lines.

In order to alleviate these problems, the ratio of the largest and smallest capacitors in the array must be decreased. For example, rather than scaling all the capacitors, their bottom plate *voltage swings* can be scaled. This can be accomplished by combining a resistor ladder with the capacitor array such that the ladder provides fractions of V_{REF} [36]. Shown in Figure 6.55 is an example where the resistor ladder generates $V_{REF}/2$, relaxing the maximum ratio required for the capacitors in the array by a factor of 2. Note that the matching of the ladder's resistors is not critical if it is used for only a few LSBs.

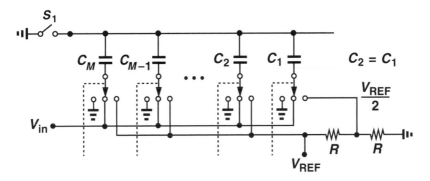

Fig. 6.55   Reduction of ratio of capacitors in charge redistribution architecture through the use of resistor ladders.

A number of successive approximation A/D converters have been realized in bipolar, CMOS, and BiCMOS technologies [34, 37, 38]. The self-calibration capabilities and density of CMOS and BiCMOS technologies have provided resolutions as high as 18 bits [38].

6.8 INTERLEAVED ARCHITECTURES

In systems where ultimate speed is the primary goal, identical A/D converters can be interleaved in time so as to achieve more parallelism than simple flash topologies. Figure 6.56(a) shows an interleaved architecture where n sample-and-hold circuits controlled by $CK_1, \ldots, CK_n$ precede n m-bit ADCs [39]. A multiplexer selects the digital output of each ADC at the proper time, providing the output corresponding to each sample. Note that $ADC_1, \ldots, ADC_n$ can employ any architecture themselves, but full flash is most commonly used to allow high a conversion rate.

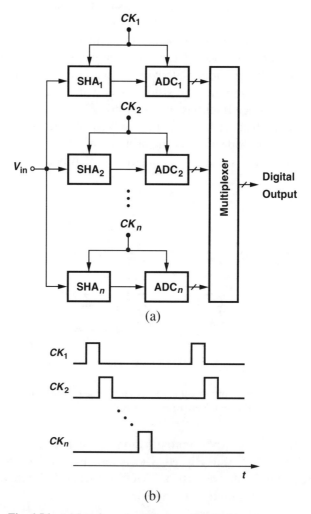

(a)

(b)

Fig. 6.56   (a) Interleaved architecture; (b) clocking sequence.

We describe the circuit's operation using the clock waveforms shown in Figure 6.56(b). When CK_j is high, SHA$_j$ acquires the analog input. When CK_j goes low, SHA$_j$ holds the instantaneous value of the input and ADC$_j$ begins to digitize that value. At the same time, CK_{j+1} goes high, allowing SHA$_{j+1}$ to acquire the next sample. The multiplexer is controlled by a combination of $CK_1, \ldots, CK_n$ such that it selects the output of ADC$_j$ when that converter has completed the digitization of a sample.

It is important to note that this architecture is advantageous over simple flash ADCs only if signal acquisition by the SHAs is sufficiently faster than A/D conversion by the ADCs.

While attaining high conversion rates, interleaved architectures suffer from accuracy degradation due to parameter *mismatches* among their constituent sampling circuits and A/D converters. In particular, mismatches in gain, differential and integral nonlinearity, timing, and offset give rise to higher noise power in the overall output. The effects of these errors in time and frequency domains have been analyzed extensively [39, 40, 41].

The potential of interleaved architectures has been demonstrated in various ADCs, most recently in an 8-bit 8-GHz system designed for real-time sampling of signals in a digital oscilloscope [42]. The system employs a total of sixteen 500-MHz bipolar flash ADCs along with a sample-and-filter technique that relaxes the bandwidths required of the preceding sampling circuits. In CMOS technology, an 8-bit 85-MHz ADC using four interleaved converters has been reported [43].

REFERENCES

[1] B. M. Gordon, "Linear Electronic Analog/Digital Conversion Architectures, Their Origins, Parameters, Limitations, and Applications," *IEEE Trans. Circuits Syst.*, vol. CAS-25, pp. 391-418, July 1978.

[2] W. Bennett, "Spectra of Quantized Signals," *Bell Syst. Tech. J.*, pp. 446-472, July 1948.

[3] R. M. Gray, "Quantized Noise Spectra," *IEEE Trans. Inf. Theory*, vol. 36, pp. 1220-1244, Nov. 1990.

[4] S. K. Tewksbury, et al., "Terminology Related to the Performance of S/H, A/D, and D/A circuits," *IEEE Trans. Circuits Syst.*, vol. CAS-25, pp. 419-426, July 1978.

[5] IEEE Standard 746-1984, "Performance Measurements of A/D and D/A Converters for PCM Television Video Circuits," IEEE, New York, 1984.

[6] T. Tsukada et al., "CMOS 8b 25MHz Flash ADC," *ISSCC Dig. Tech. Pap.*, pp. 34-35, Feb. 1985.

[7] P. Vorenkamp and J. P. M. Verdaasdonk, "A 10 b 50 Ms/s Pipelined ADC," *ISSCC Dig. Tech. Pap.*, pp. 32-33, Feb. 1992.

[8] R. Jewett, J. J. Corcoran, and G. Steinbach, "A 12 b 20 MS/sec Ripple-Through ADC," *ISSCC Dig. Tech. Pap.*, pp. 34-35, Feb. 1992.

[9] T. Shimizu et al., "A 10-bit 20-MHz Two-Step Parallel A/D Converter with Internal S/H," *IEEE J. Solid-State Circuits*, vol. SC-24, pp. 13-20, Feb. 1989.

[10] A. G. Dingwall and V. Zazzu, "An 8-MHz CMOS Subranging 8-Bit A/D Converter," *IEEE J. Solid-State Circuits*, vol. SC-20, pp. 1138-1143, Dec. 1985.

[11] N. Fukushima et al., "A CMOS 40 MHz 8 b 105 mW Two-Step ADC," *ISSCC Dig. Tech. Pap.*, pp. 14-15, Feb. 1989.

[12] B. Peetz, B. D. Hamilton, and J. Kang, "An 8-Bit 250 Megasample/sec A/D Converter," *IEEE J. Solid-State Circuits*, vol. SC-21, pp. 997-1002, Dec. 1986.

[13] C. W. Mangelsdorf et al., "A 400-MHz Input Flash Converter with Error Correction," *IEEE J. Solid-State Circuits*, vol. SC-25, pp. 184-191, Feb. 1990.

[14] Y. Akazawa et al., "A 400 MSPS 8 b Flash AD Conversion LSI," *ISSCC Dig. Tech. Pap.*, pp. 98-99, Feb. 1987.

[15] V. E. Garuts et al., "A Dual 4-Bit 1.5 GS/s Analog-to-Digital Converter," *Proc. BCTM*, pp. 141-144, Sept. 1988.

[16] B. Zojer, R. Petschacher, and W. A. Luschnig, "A 6-Bit 200-MHz Full Nyquist A/D Converter," *IEEE J. Solid-State Circuits*, vol. SC-20, pp. 780-786, June 1985.

[17] B. S. Song, S. H. Lee, and M. F. Tompsett, " A 10-Bit 15-MHz CMOS Recycling Two-Step A/D Converter," *IEEE J. Solid-State Circuits*, vol. SC-25, pp. 1328-1338, Dec. 1990.

[18] K. Tsugaru et al., "A 10 Bit 40 MHz ADC Using 0.8 μm BiCMOS Technology," *Proc. BCTM*, pp. 48-51, Sept. 1989.

[19] M. Steyaert, R. Roovers, and J. Cranickx, "A 100 MHz 8 Bit CMOS Interpolating A/D Converter," *Proc. CICC*, pp. 28.1.1-28.1.4, May 1993.

[20] K. Kusumoto, A. Matsuzawa, and K. Murata, "A 10-b 20-MHz 30-mW Pipelined Interpolating CMOS ADC," *IEEE J. Solid-State Circuits*, vol. SC-28, pp. 1200-1206, Dec. 1993.

[21] C. Lane, "A 10-Bit 60 MSPS Flash ADC," *Proc. BCTM*, pp. 44-47, Sept. 1989.

[22] R. E. J. van de Grift, I. W. J. M. Rutten, and M. van der Veen, "An 8-Bit Video ADC Incorporating Folding and Interpolation Techniques," *IEEE J. Solid-State Circuits*, vol. SC-22, pp. 944-953, Dec. 1987.

[23] H. Kimura et al., "A 10-b 300-MHz Interpolated Parallel A/D Converter," *IEEE J. Solid-State Circuits*, vol. SC-28, pp. 438-446, Apr. 1993.

[24] J. van Valburg and R. J. van de Plassche, "An 8-b 650-MHz Folding ADC," *IEEE J. Solid-State Circuits*, vol. SC-27, pp. 1662-1666, Dec. 1992.

[25] R. J. van de Plassche, "An 8-Bit 100-MHz Full-Nyquist Analog-to-Digital Converter," *IEEE J. Solid-State Circuits*, vol. SC-27, pp. 1334-1344, Dec. 1988.

[26] U. Fiedler and D. Seitzer, "A High-Speed 8 Bit A/D Converter Based on a Gary-Code Multiple Folding Circuit," *IEEE J. Solid-State Circuits*, vol. SC-14, pp. 547-551, June 1979.

[27] R. J. van de Plassche and R. E. J. van der Grift, "A High-Speed 7 Bit A/D Converter," *IEEE J. Solid-State Circuits*, vol. SC-14, pp. 938-943, Dec. 1979.

[28] R. E. J. van de Grift and R. J. van de Plassche, "A Monolithic 8-Bit Video A/D Converter," *IEEE J. Solid-State Circuits*, vol. SC-19, pp. 374-378, June 1984.

[29] W. T. Colleran and A. A. Abidi, "A 10-b 75-MHz Two-Step Pipelined Bipolar A/D Converter," *IEEE J. Solid-State Circuits*, vol. SC-28, pp. 1187-1199, Dec. 1993.

[30] M. Yotsuyanagi, T. Etoh, and K. Hirata, "A 10 Bit 50 MHz Pipelined CMOS A/D Converter with S/H," *IEEE J. Solid-State Circuits*, vol. SC-28, pp. 292-300, March 1993.

[31] B. S. Song, M. F. Tompsett, and K. R. Lakshmikumar, "A 12-Bit 1-Msample/s Capacitor-Averaging Pipelined A/D Converter," *IEEE J. Solid-State Circuits*, vol. SC-23, pp. 1324-1333, Dec. 1988.

[32] Y. M. Lin, B. S. Kim, and P. R. Gray, "A 13-b 2.5-MHz Self-Calibrated Pipelined A/D Converter in 3-μm CMOS," *IEEE J. Solid-State Circuits*, vol. SC-26, pp. 628-636, April 1991.

[33] K. Sone, Y. Nishida, and N. Nakadai, "A 10-b 100-Msample/sec Sub-ranging BiCMOS ADC," *IEEE J. Solid-State Circuits*, vol. SC-28, pp. 1187-1199, Dec. 1993.

[34] K. Bacrania, "A 12-Bit Successive-Approximation-Type ADC with Digital Error Correction," *IEEE J. Solid-State Circuits*, vol. SC-21, pp. 1016-1025, Dec. 1986.

[35] J. L. McCreary and P. R. Gray, "All-MOS Charge Redistribution Analog-to-Digital Conversion Techniques—Part I," *IEEE J. Solid-State Circuits*, vol. SC-10, pp. 371-379, Dec. 1975.

[36] B. Fotouhi and D. A. Hodges, "High-Resolution A/D Conversion in MOS/LSI," *IEEE J. Solid-State Circuits*, vol. SC-14, pp. 920-926, Dec. 1979.

[37] R. K. Hester et al., "Fully Differential ADC with Rail-to-Rail Common Mode Range and Nonlinear Capacitor Compensation," *IEEE J. Solid-State Circuits*, vol. SC-25, pp. 173-183, Feb. 1990.

[38] G. A. Miller, "An 18 b 10 μs Self-Calibrating ADC," *ISSCC Dig. Tech. Pap.*, pp. 168-169, Feb. 1990.

[39] W. Black and D. A. Hodges, "Time Interleaved Converter Arrays,"*IEEE J. Solid-State Circuits*, vol. SC-15, pp. 1022-1029, Dec. 1980.

[40] A. Montijo and K. Rush, "Accuracy in Interleaved ADC Systems," *Hewlett-Packard J.*, pp. 38-46, Oct. 1993.

[41] Y. C. Jenq, "Digital Spectra of Nonuniformly Sampled Signals: Fundamentals and High-Speed Waveform Digitizers," *IEEE Trans. Instrum. Meas.*, vol. 37, pp. 245-251, June 1988.

[42] M. T. McTigue and P. J. Byrne, "An 8-Gigasample/sec 8-Bit Data Acquisition System for a Sampling Digital Oscilloscope," *Hewlett-Packard J.*, pp. 11-23, Oct. 1993.

[43] C. S. G. Conroy, D. W. Cline, and P. R. Gray, "An 8-b 85-MS/s Parallel Pipelined A/D Converter in 1-μm CMOS," *IEEE J. Solid-State Circuits*, vol. SC-28, pp. 447-454, April 1993.

7

Building Blocks
of Data Conversion Systems

The design of data conversion systems deals with both architectural issues and circuit level considerations. The choice of an architecture is influenced not only by its viability in a given technology but also by the performance of its constituent building blocks. In fact, the trade-offs that exist at each level of abstraction often mandate a great deal of iteration between architecture and circuit design.

In this chapter, we describe the design of building blocks for data acquisition and conversion. These include open-loop amplifiers, operational amplifiers, and comparators. Issues related to the speed and precision of each circuit are detailed and the impact of each building block's performance on that of data conversion systems is discussed.

7.1 AMPLIFIERS

Amplifiers are integral components of analog signal processing systems. In data acquisition, they are utilized in sampling circuits, subtractors, gain stages, and pipelined circuits, with each application imposing a different set of requirements on their speed, gain, linearity, noise, dynamic range, and power dissipation. Here, we study both open-loop and closed-loop amplifiers.

7.1.1 Open-Loop Amplifiers

Open-loop amplifiers have been popular in bipolar technology for two reasons. First, many bipolar processes do not provide high-speed (vertical) pnp transistors and hence make it difficult to design high-gain amplifiers (which usually require high-impedance loads). Thus, only low to moderate gains are feasible, implying that the benefits resulting from feedback around amplifiers are quite limited here. Second, open-loop amplifiers generally exhibit faster settling times than their closed-loop counterparts.

Despite these advantages, open-loop amplifiers must deal with nonidealities such as nonlinearity and gain error, which would be suppressed in closed-loop configurations. Consequently, a variety of correction techniques have been devised to improve the performance of these amplifiers without employing global feedback. While linearity is crucial in all circuits, gain error becomes important only in interstage amplifiers [1] and certain sampling circuits [2]. In order to understand the linearity issues, we first analyze a simple bipolar differential pair.

Nonlinearity in a Bipolar Differential Pair. For a simple bipolar differential pair such as that shown in Figure 7.1(a), the large-signal output differential voltage can be expressed as [3]

$$V_{od} = R_C I_{EE} \tanh \frac{V_{id}}{2V_T}, \tag{7.1}$$

where $V_T = kT/q$, $R_C = R_{C1} = R_{C2}$, and the effect of β and base and emitter resistance is neglected. The hyperbolic tangent term in (7.1) approaches a saturation level of ± 1 as $|V_{id}|$ exceeds approximately $6V_T$. Thus, the input

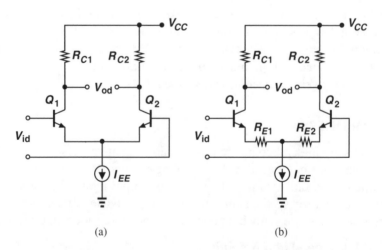

(a) (b)

Fig. 7.1 (a) Simple bipolar differential pair; (b) emitter-degenerated differential pair.

linear range is quite limited. For example, since for $\epsilon \ll 1$, $\tanh \epsilon \approx \epsilon - \epsilon^3/3$, (7.1) can be simplified if $V_{id} \ll 2V_T$:

$$V_{od} \approx R_C I_{EE} \left(\frac{V_{id}}{2V_T} - \frac{V_{id}^3}{24V_T^3} \right). \tag{7.2}$$

The second term in the parentheses represents nonlinearity in the input/output characteristic and is greater than 1% of the first term for a V_{id} as small as $0.35V_T$.

From a small-signal point of view, the gain of the circuit, $A_v = V_{od}/V_{id}$, can be expressed as

$$A_v = \frac{2R_C}{g_{m1}^{-1} + g_{m2}^{-1}} \tag{7.3}$$

$$= \frac{2R_C}{V_T} \frac{I_{C1}I_{C2}}{I_{C1} + I_{C2}} \tag{7.4}$$

$$= \frac{2R_C}{V_T} \frac{I_{C1}I_{C2}}{I_{EE}}, \tag{7.5}$$

suggesting that the gain varies as a function of the differential input voltage because I_{C1} and I_{C2} vary. Note that A_v also depends on temperature.

The nonlinearity expressed by (7.3) arises because the denominator is a current-dependent impedance. Therefore, the nonlinearity can be reduced by either suppressing this dependence in the denominator or creating the same dependence in the numerator. The first approach leads to emitter degeneration, and the second to logarithmic loads.

Linearized Bipolar Differential Pair.   Figure 7.1(b) illustrates an emitter-degenerated differential pair. Here, the small-signal gain is

$$A_v = \frac{2R_C}{g_{m1}^{-1} + g_{m2}^{-1} + 2R_E}, \tag{7.6}$$

indicating that A_v becomes less input- and temperature-dependent as $2R_E$ ($= 2R_{E1} = 2R_{E2}$) becomes much greater than $g_{m1}^{-1} + g_{m2}^{-1}$. (This is equivalent to making $I_{EE}R_E$ much greater than V_T.) The improvement in linearity, however, is attained at the cost of lowering the gain or allowing a larger voltage drop across the collector resistors. The emitter resistors also raise the input thermal noise and offset voltage.

Shown in Figure 7.2(a) is a differential pair employing diode-connected transistors as logarithmic loads. The small-signal gain of this circuit is

$$A_v = \frac{g_{m3}^{-1} + g_{m4}^{-1}}{g_{m1}^{-1} + g_{m2}^{-1}} \tag{7.7}$$

$$= \frac{V_T(I_{C3} + I_{C4})/(I_{C3}I_{C4})}{V_T(I_{C1} + I_{C2})/(I_{C1}I_{C2})} \tag{7.8}$$

$$\approx 1. \tag{7.9}$$

It follows from (7.9) that the gain of this circuit remains independent of the bias current—and hence the input voltage—and is close to unity. In practice, however, nonidealities such as finite β, finite Early voltage, and degradation of β at low currents limit the usable input range of the circuit to approximately $10V_T$.

The performance of the two circuits discussed above can be substantially improved by combining the two techniques, as depicted in Figure 7.2(b). Here, the gain can be expressed as

$$A_v = \frac{g_{m3}^{-1} + R_{C1} + g_{m4}^{-1} + R_{C2}}{g_{m1}^{-1} + R_{E1} + g_{m2}^{-1} + R_{E2}}, \tag{7.10}$$

which for $R_{E1} = R_{E2} = R_{C1} = R_{C2} = R$, $g_{m1} = g_{m3}$, and $g_{m2} = g_{m4}$ becomes equal to unity.

In the circuit of Figure 7.2(b), the voltage drop across R_{E1} and R_{E2} severely limits the input and output voltage swings if a high linearity is required. To alleviate this problem, I_{EE} can be split into two equal current sources that directly flow from Q_1 and Q_2. In such a configuration, shown in Figure 7.2(c), the current flowing through $R_{E1} + R_{E2}$ is determined only by the differential input voltage, allowing emitters of Q_1 and Q_2 to have voltages closer to the negative rail.

The circuits of Figure 7.2 suffer from gain error introduced by the finite (ac) β of devices. Starting with (7.10), the reader can easily show that the gain error is equal to $\beta/(\beta + 1)$. For a typical β of 100, this yields a gain error of 1%.

This error can be significantly reduced by adding resistors with proper values in series with bases of Q_3 and Q_4 (Figure 7.3) [1]. Assuming an infinite Early voltage for all the transistors and noting that the impedance seen looking into the emitter of Q_3 or Q_4 is equal to $(R_B + r_\pi)/(\beta + 1)$, where $R_B = R_{B1} = R_{B2}$, the reader can show that the small-signal gain of

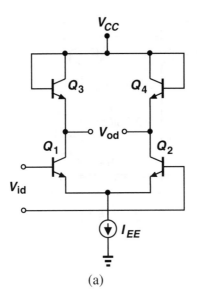

(a)

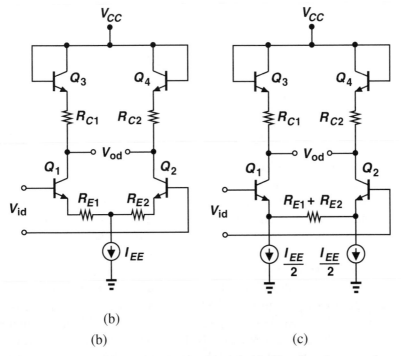

(b)

(b)   (c)

Fig. 7.2   (a) Differential pair with logarithmic loads; (b) emitter-degenerated
differential pair with logarithmic loads; (c) circuit of (b) with split
emitter currents.

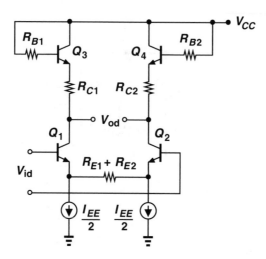

Fig. 7.3   Differential pair with resistors R_{B1} and R_{B2} added to correct for finite β.

this circuit is expressed as

$$A_v = \frac{\beta}{\beta+1} \frac{R + \dfrac{R_B + r_\pi}{\beta+1}}{R + \dfrac{r_\pi}{\beta+1}}, \tag{7.11}$$

where $R = R_{E1} = R_{E2} = R_{C1} = R_{C2}$. In order for A_v to be identical with unity,

$$R_B = \frac{\beta+1}{\beta}R + \frac{1}{\beta}r_\pi. \tag{7.12}$$

As an easy, reliable choice, $R_B \approx R$, for which the gain error becomes proportional to $1/\beta^2$ rather than $1/\beta$.

In addition to the techniques described above, other methods of improving the linearity and gain error of differential pairs have been proposed [4, 5]. A common drawback of most of these configurations is their limited input and output voltage swings due to stacked devices and voltage drops across emitter and collector resistors. This is particularly troublesome for high resolutions because these voltage drops must be large to ensure high linearity, while the input and output swings must also be large to provide a wide dynamic range. For example, when designed for a gain of 2, the circuit of Figure 7.3 requires ±5-V supplies to attain a 2-V swing (a 4-V differential) and 12-bit linearity and gain precision [1]. The trade-off between the linearity and dynamic range of such amplifiers makes them less attractive for higher resolutions, especially if the supply voltage is limited to 5 V.

MOS Differential Pair.   The precision techniques described above for open-loop bipolar circuits do not achieve the same level of performance in CMOS. For example, consider the CMOS counterpart of Figure 7.2(a), shown in Figure 7.4(a). If the body of M_3 and M_4 is connected to ground, the small-signal gain is

$$A_v = \frac{(g_{m3} + g_{mb3})^{-1} + (g_{m4} + g_{mb4})^{-1}}{g_{m1}^{-1} + g_{m2}^{-1}}, \tag{7.13}$$

where g_{mj} denotes the transconductance of M_j and g_{mbj} represents the small-signal impedance due to the body effect [3]. The additional term g_{mbj} in this expression introduces gain error (and nonlinearity) and is difficult to cancel.

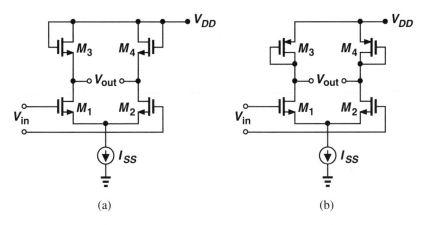

(a)   (b)

Fig. 7.4   MOS differential pair with diode-connected (a) NMOS and (b) PMOS loads.

In order to remove the error due to the body effect, the load devices can be implemented with PMOS transistors, as depicted in Figure 7.4(b), where $A_v = (g_{m3}^{-1} + g_{m4}^{-1})/(g_{m1}^{-1} + g_{m2}^{-1})$. However, the gain of this circuit is not easily defined in modern MOS technologies. To understand why, first assume M_1-M_4 are long-channel, square-law devices having equal lengths and $g_m = \sqrt{2I_D W \mu C_{ox}/L}$. Thus,

$$A_v = \sqrt{\frac{W_{12}\mu_n}{W_{34}\mu_p}}, \tag{7.14}$$

where $W_{12} = W_1 = W_2$ and $W_{34} = W_3 = W_4$. This equation indicates that the gain can be precisely set by the ratio $(W_{12}\mu_n)/(W_{34}\mu_p)$. For short-channel devices, however, carrier velocity saturation is significant and the drain current of a MOSFET is approximately equal to

$$I_D \approx W C_{ox} (V_{GS} - V_{TH}) v_{sat}, \tag{7.15}$$

where v_{sat} is the saturation velocity of carriers [6]. The transconductance of such a device can be expressed as

$$g_m = \frac{\partial I_D}{\partial V_{GS}} \tag{7.16}$$

$$= W C_{ox} v_{sat}, \tag{7.17}$$

which is independent of current and device length. Thus, the small-signal gain of the circuit in Figure 7.4(b) will be

$$A_v = \frac{W_{12} v_{sat,n}}{W_{34} v_{sat,p}}. \tag{7.18}$$

In a typical technology with $L < 1\,\mu\text{m}$, devices are not fully velocity-saturated but they deviate substantially from square-law behavior. As a result, the actual gain has a magnitude between $\sqrt{(W_{12}\mu_n)/(W_{34}\mu_p)}$ and $(W_{12}\mu_n)/(W_{34}\mu_p)$, thereby making it difficult to predict its exact value.

Another point of contrast between bipolar and CMOS amplifiers is evident in follower circuits. While emitter followers are frequently used in analog (and digital) design, source followers have not been as popular for several reasons. First, unless wide devices or high-bias currents are used, the output impedance of a source follower is roughly the same as that of circuits such as in Figure 7.4, indicating source followers are not efficient drivers. Second, the gate-source level shift (typically greater than 1 V) often limits the voltage swings and hence the dynamic range. Third, if the body of a source follower is tied to a constant potential, the body effect introduces substantial nonlinearity and gain error.

For these reasons, precision amplification in CMOS technology is often performed using closed-loop configurations.

7.1.2 Closed-Loop Amplifiers

Closed-loop amplifiers are frequently used in precision processing of analog signals because negative feedback can yield accurate closed-loop gain and high linearity. Additionally, negative feedback can provide virtual ground nodes, a necessity when precise amounts of charge must be transferred from one capacitor to another. For these reasons, operations such as sampling and subtraction usually employ closed-loop circuits for precisions above roughly 10 bits.

The precision resulting from negative feedback of course depends on the gain and linearity of the *open-loop* circuit. In particular, an important challenge in the design of closed-loop circuits is achieving a high open-loop gain while maintaining reasonable speed, voltage swings, and power dissipation.

While traditionally phase margin and unity-gain bandwidth are used to predict the speed of closed-loop circuits, it is extremely important to realize that these parameters are *small-signal* quantities unable to represent the large-signal behavior of such circuits. Thus, to obtain a more realistic view, the large-signal time response of the circuit (in a closed loop), including both slew rate and settling time, must be examined.

Unity-Gain Buffer with Local Feedback.   Figure 7.5 shows a BiC-MOS unity-gain buffer, consisting of a differential pair Q_1-Q_2 with active load M_1-M_2 and an emitter follower Q_3 in a local feedback loop. The small-signal open-loop gain of this circuit is

$$A_{vo} \approx g_{m12}(r_{OQ2} \| r_{OM2}), \tag{7.19}$$

where g_{m12} is the transconductance of Q_1 and Q_2 and r_{OQ2} and r_{OM2} represent the output impedance of Q_2 and M_2, respectively. The open-loop input and output impedance of the amplifier are

$$R_{\text{in},o} \approx 2r_{\pi12} \tag{7.20}$$

$$R_{\text{out},o} \approx \frac{r_{OQ2} \| r_{OM2}}{\beta + 1} + \frac{1}{g_{m3}}. \tag{7.21}$$

In the closed-loop circuit, the gain is $A_{vc} = A_{vo}/(1 + A_{vo})$, and the equivalent input and output impedance are $R_{\text{in},c} \approx R_{\text{in},o} \cdot A_{vo}$ and $R_{\text{out},c} \approx R_{\text{out},o}/A_{vo}$. Thus,

$$A_{vc} \approx 1 - \frac{1}{g_{m12}(r_{OQ2} \| r_{OM2})} \tag{7.22}$$

$$R_{\text{in},c} \approx 2\beta(r_{OQ2} \| r_{OM2}) \tag{7.23}$$

$$R_{\text{out},c} \approx \frac{1}{\beta g_{m12}}. \tag{7.24}$$

It follows from the above equations that the buffer achieves a high input impedance, a low output impedance, and a gain error of $[g_{m12}(r_{OQ2} \| r_{OM2})]^{-1}$ (typically a few tenths of a percent). The maximum input and output voltage swings are given by the difference between the supply voltage and the sum of $V_{\text{DS},2}$, $V_{\text{BE},3}$, $V_{\text{BE},2}$, and the voltage required across current source I_{EE} (assuming $V_{\text{DS},2} + V_{\text{BE},3} \geq V_{\text{GS},1}$). These swings are approximately equal to 2.5 V in a 5-V system.

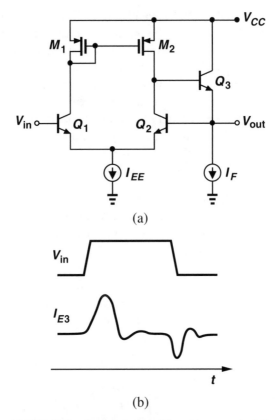

Fig. 7.5 (a) BiCMOS unity-gain buffer; (b) emitter current of Q_3 when the circuit drives a large load capacitance.

Despite these features, the buffer of Figure 7.5 suffers from a number of speed-related drawbacks. In addition to a dominant pole at the collector of Q_2, the circuit has nondominant poles at the collector of Q_1 and emitter of Q_3. The latter is given by the output impedance and the total capacitance seen at that node and may substantially degrade the settling behavior if the buffer drives a large load capacitance. Another important effect, often ignored in small-signal analysis, is the *variation* of this pole during large-signal transients at the output [7]. Since I_F is constant, during transients the emitter current of Q_3 must change to allow the load capacitance to charge or discharge. For example, during a positive voltage excursion at the output, Q_3 carries not only I_F but also the load capacitance current. Figure 7.5(b) depicts the emitter current of Q_3 for both positive and negative voltage excursions, indicating that the pole magnitude first changes significantly and then requires a long time to settle to its original value.

In order to overcome this problem, I_F can be controlled such that Q_3 always carries a relatively constant current; i.e., when the load capacitance current increases, I_F decreases, and vice versa [7]. Shown in Figure 7.6 is a possible implementation of this concept, where amplifier A adjusts I_F according to the difference between the input and output voltages, making I_F an "active pull-down" device. This circuit can also be viewed as two parallel amplifiers that drive the load in a push-pull fashion. Of course A must be sufficiently fast so as not to introduce additional settling components at the output.

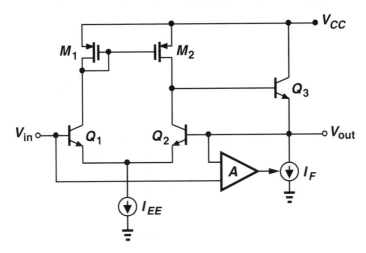

Fig. 7.6   Additional feedback applied to the circuit of Figure 7.5 to maintain I_{E3} constant.

The pole formed at the collector of Q_2 often does not provide sufficient roll-off in the gain, causing significant ringing in the step response, especially for a large load capacitance. To alleviate the problem, this pole can be brought closer to the origin by adding more capacitance from the collector of Q_2 to ground [Figure 7.7(a)]. Alternatively, pole-splitting techniques [3] such as that in Figure 7.7(b) can be used to both move this pole further from the origin and create another dominant pole at the base of Q_2.

The topology of Figure 7.5 must be modified if complementary devices or bipolar transistors are not available. Figure 7.8 shows two variants in all-npn bipolar and pure CMOS technologies. Since the open-loop gain of these circuits is about an order of magnitude less than that of Figure 7.5, their gain error is quite significant. The nonlinearity of the bipolar buffer is also substantial because its loop gain is relatively small.

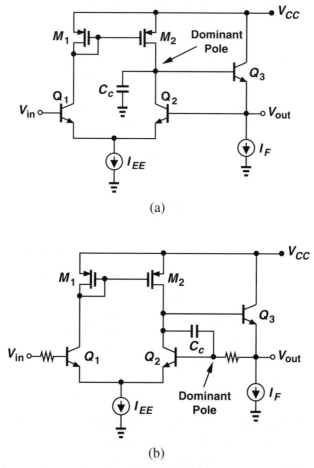

(a)

(b)

Fig. 7.7   Compensation of unity-gain buffer by (a) lowering the magnitude of the dominant pole and (b) pole splitting.

7.1.3 Operational Amplifiers

The large gain and high linearity that can be achieved in operational amplifiers often prove crucial in signal processing. For example, many of the SHA and ADC architectures described in Chapters 3 and 6 require such amplifiers so as to attain high resolutions.

While op amps traditionally were designed to have a gain of several hundred thousand and provide a relatively low output impedance (even in open-loop configuration), they have evolved into different topologies as supply voltages have scaled down and CMOS devices have become popular in analog design. Various trade-offs among gain, speed, dynamic range, and

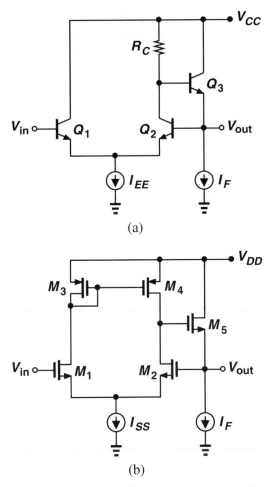

Fig. 7.8 Unity-gain buffers in (a) all-npn bipolar and (b) CMOS technologies.

power dissipation of op amps prohibit an arbitrary choice of any of these parameters, thereby mandating custom design for every system.

Since dynamic range is a crucial parameter in high-resolution systems—and is limited by the supply voltage—it cannot be simply compromised for other parameters. In particular, op amps are usually designed to achieve maximum input and output voltage swings and low input noise so as to allow a wide dynamic range. In this respect, low-voltage designs usually avoid emitter followers or source followers at the output because of the additional headroom consumed by these circuits. However, such designs cannot drive low-*resistance* loads because their open-loop gain falls sharply.

Another attribute of modern op amps is that their open-loop gain is commensurate with the maximum allowable gain error of the closed-loop circuit. Resulting from the trade-offs mentioned above, this choice of gain assumes that the nonlinearity is also suppressed to sufficiently low levels, e.g., roughly the same as the gain error. This assumption is reasonable in most topologies and simplifies the design if the gain error and nonlinearity are equally important (e.g., in interstage SHAs of pipelined ADCs).

Two-Stage Op Amps. The two-stage topology has been successfully used in many general-purpose, high-gain op amps [8]. Shown in Figure 7.9, this configuration consists of a transconductance stage G_{m1}, a voltage amplifier A_1, and a Miller compensation capacitor C_C placed around the amplifier. The small-signal open-loop gain of the circuit is

$$A_{vo} = G_{m1} R_1 A_1, \qquad (7.25)$$

where R_1 is the resistance seen at the output of G_{m1} and equals the parallel combination of the output resistance of G_{m1} and the input resistance of A_1. Since the overall gain is equal to the *product* of the voltage gain of each stage, it can be very high. The -3-dB open-loop bandwidth of the circuit is

$$BW_o \approx \frac{1}{2\pi R_1 A_1 C_C}. \qquad (7.26)$$

It follows from (7.25) and (7.26) that the gain-bandwidth product of the op amp is

$$A_{vo} \cdot BW_o \approx \frac{G_{m1}}{2\pi C_C}. \qquad (7.27)$$

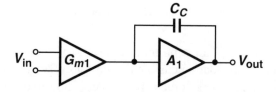

Fig. 7.9 Two-stage op amp configuration.

The principal drawback of this architecture stems from the nondominant pole formed by the output impedance of A_1 and the load capacitance. This pole severely degrades the settling behavior because, as mentioned above, amplifier A_1 usually employs no source or emitter followers at its output and hence exhibits a relatively high output impedance. Since G_{m1} and A_1 typically have additional poles, the op amp bandwidth must be drastically reduced so as to avoid underdamped settling at the output. Furthermore, the feedforward of signal through C_C gives a zero in the right half-plane, often requiring other circuit techniques to ensure stability [9, 10].

The two-step nature of this configuration offers an important feature that can be exploited if maximum dynamic range is the primary concern: the G_m stage can provide a high gain, while amplifier A_1 is designed for nearly rail-to-rail output swings [11]. This allocation of gain and voltage swing is not possible in op amp topologies that provide the entire gain in one stage.

Cascode Op Amps.   Since most of the op amps employed in data acquisition systems need not drive resistive loads, they can be designed without concern for their open-loop output resistance. As such, these op amps can have a very *high* output resistance, i.e., designed as transconductance amplifiers. The advantage of this approach is that a relatively high voltage gain can be obtained in one stage.

Figure 7.10(a) depicts a CMOS cascode op amp with single-ended output [3]. The small-signal open-loop voltage gain of the circuit is equal to $g_{m12}R_O$, where R_O is the impedance seen at the drain of M_4; i.e., $R_O \approx (g_{m4}r_{O4}r_{O2})\|(g_{m6}r_{O6}r_{O8})$. This gain is typically around a few thousand, and the circuit offers the linearity and gain error required for resolutions up to 10 bits [12]. The dominant pole is formed at the output node by the load capacitance.

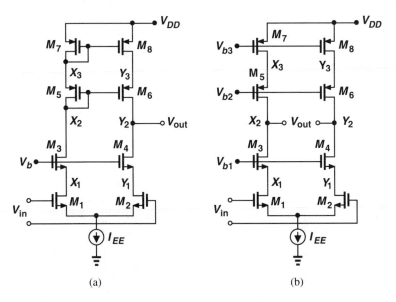

(a)   (b)

Fig. 7.10   (a) Single-ended cascode op amp; (b) fully differential cascode op amp (common-mode feedback not shown).

This topology suffers from a number of trade-offs among its gain, dynamic range, slew rate, and settling time. Since the circuit employs a stack of five devices (including the current source), its output voltage swing is quite limited. Thus, to achieve both a large output swing and a high gain, the devices must be wide so that their $V_{GS} - V_{TH}$ is minimized and their transconductance maximized. On the other hand, to attain a high slew rate, the bias currents must be made large, lowering the open-loop gain unless the transistors are made longer. The resulting size of the devices yields large input capacitance and large parasitics at nodes X_1-X_3 and Y_1-Y_3, thereby reducing the magnitude of the poles formed at these nodes and degrading the settling characteristics. In particular, the pole associated with the mirror device M_7 typically becomes the primary nondominant pole, degrading the phase margin and limiting the unity-gain bandwidth. This pole moves toward the origin as the design maximizes the output voltage swing by increasing the width of the transistors. It can be easily shown that, for a given $V_{GS,7}$, the magnitude of the pole at node X_3 decreases as the drain current and width of M_7 increase together.

To alleviate some of these problems, a fully differential topology such as that of Figure 7.10(b) can be used. In this circuit, as is often said, signals do not propagate through PMOS devices. This is a relatively accurate statement because the only signals that appear at nodes X_3 and Y_3 result from reverse characteristics of common-gate devices M_5-M_6, i.e., their finite output impedance. While the parasitic capacitance at these nodes shunts the output impedance of M_7 and M_8 (thus producing a pole), the impedance seen looking into the drain of M_5 and M_6 has only one pole (even in the presence of load capacitance). We calculate the impedance looking into the drain of M_5 by noting that

$$Z_{O5} \approx g_{m5}r_{O5}(r_{O7}||\frac{1}{C_7s}), \qquad (7.28)$$

where C_7 is the total capacitance at the drain of M_7. This expression can be written as

$$Z_{O5} \approx (g_{m5}r_{O5}r_{O7})||(\frac{g_{m5}r_{O5}}{C_7s}), \qquad (7.29)$$

suggesting that Z_{O5} is equivalent to the parallel combination of a resistor equal to $g_{m5}r_{O5}r_{O7}$ and a capacitor equal to $C_7/(g_{m5}r_{O5})$.

Since the load capacitance C_L simply appears in parallel with this combination, we can write

$$Z_{O5}||\frac{1}{C_Ls} = (g_{m5}r_{O5}r_{O7})||\frac{1}{(C_L + \dfrac{C_7}{g_{m5}r_{O5}})s}. \qquad (7.30)$$

It follows from (7.30) that the capacitance at nodes X_3 and Y_3 is divided by the intrinsic gain of M_5 and M_6 and "absorbed" by the load capacitance. In reality, the approximations made in arriving at (7.28) (e.g., the assumption that $g_{m5}r_{O5} \gg 1$) ignore some higher-order terms in s that could introduce pole-zero doublets. But, the effect of these terms is usually negligible.

With the mirror pole removed, the nondominant pole of the circuit in Figure 7.10(b) arises at nodes X_1 and Y_1 and is determined by the transconductance of M_3 and M_4 and the total capacitance at these nodes. In a typical design, this pole is several times higher than the mirror pole of Figure 7.10(a).

The fully differential topology of Figure 7.10(b) requires a common-mode feedback network so as to remain in its high-gain region. This subject is discussed in Section 7.1.5.

The circuit of Figure 7.10(b) still suffers from the same dynamic range trade-offs described for the circuit of Figure 7.10(a). In order to increase the input and output swings, the cascode topology can be "folded," as illustrated in Figure 7.11. Here, the input stage has three stacked devices and the output stage four, giving larger input and output swings than the circuits of Figure 7.10. However, PMOS devices M_5 and M_6 are in the signal path, creating a nondominant pole at the folding points X_1 and Y_1. Given by the transconductance of M_5 and M_6 and the total capacitance at these nodes, this pole usually determines the phase margin and maximum bandwidth of the op amp. If M_1 and M_2 are replaced with a PMOS differential pair with their drains connected to the source of M_7 and M_8, then the pole associated with the folding points is given by the transconductance of NMOS devices M_7 and M_8 and the total capacitance at their source, a potentially higher magnitude than that in Figure 7.11. However, the open-loop gain tends to be lower because of input PMOS transistors.

The circuits of Figures 7.10 and 7.11 exhibit an interesting trade-off between their input-referred noise and output voltage swings. In Figure 7.10, for example, since the contribution of M_7 and M_8 to the input-referred rms noise increases with the square root of their transconductance [3] and since $g_m \approx 2I_D/(V_{GS} - V_{TH})$, the noise increases if $V_{GS} - V_{TH}$ of these transistors is minimized to allow a large output swing. Note that the folded cascode exhibits more noise than the unfolded counterpart because of the contribution of four devices, M_3, M_4, M_9, and M_{10} in Figure 7.11.

The op amp topologies described above can also be implemented in BiCMOS or complementary bipolar technologies. In particular, the folded-cascode configuration has a BiCMOS counterpart, shown in Figure 7.12 [7]. In this circuit, the pole at nodes X_1 and Y_1 is on the order of the f_T of the bipolar

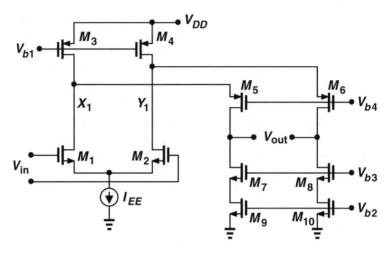

Fig. 7.11   CMOS differential folded-cascode op amp (common-mode feedback not shown).

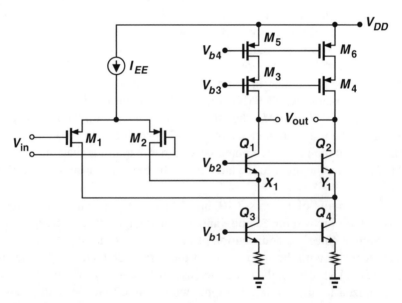

Fig. 7.12   BiCMOS folded-cascode differential op amp (common-mode feedback not shown).

transistors Q_1 and Q_2, yielding a faster settling time than a pure CMOS implementation.

While bipolar transistors can provide a higher gain and lower noise than CMOS devices, most BiCMOS op amps still avoid them in the input stage.

The reasons why are as follows. First, the input bias current introduces droop and offset voltage in switched-capacitor circuits and is difficult to cancel. Second, if the folded-cascode topology incorporates an npn input pair, it will inevitably have PMOS devices in the signal path, thus suffering from a significant nondominant pole and hence long settling times.

7.1.4 Gain Boosting Techniques

The bandwidth limitations of two-stage op amps and the limited gain achievable in folded-cascode configurations have motivated the invention of gain-boosting techniques [13, 14]. These techniques are usually applied to cascode op amps to increase their gain with little degradation in speed.

To understand the principle of gain boosting, consider the cascode circuit of Figure 7.13(a). Here, if I_1 is an ideal current source, the small-signal gain is approximately equal to $g_{m1}g_{m2}r_{O1}r_{O2}$, i.e., the transconductance of the input transistor multiplied by the impedance seen at the output node. The output impedance and hence the gain can be increased by stacking more devices in the cascode configuration, but at the cost of reduced output swings. Now suppose, as shown in Figure 7.13(b), M_2 is placed in a feedback loop that senses its drain current and adjusts its gate voltage so as to minimize *variations* in the drain current. In other words, if a change in V_{out} tends to vary V_X through the output impedance of M_2, then A_0 varies the gate voltage of M_2 in such a way to minimize the change in V_X. Since the feedback loop senses the output current, the output impedance of the circuit is boosted by approximately A_0, yielding $A_{v,boost} = A_0 g_{m1}g_{m2}r_{O1}r_{O2}$.

It is interesting to note that while in a two-stage op amp the "entire" signal must propagate through both stages, in the circuit of Figure 7.13(b) only an *error* signal is processed by A_0. This in turn means the settling times are much faster in the latter than in the former. Also, A_0 need not have large output swings or a high slew rate and hence can be optimized for small-signal gain and bandwidth, providing a high overall gain and fast settling.

The amplifier topology used for A_0 depends on the overall gain requirement and can be as simple as a common-source stage, as shown in Figure 7.13(c) [14]. In a typical design, the common-source stage, consisting of M_3 and I_2, boosts the gain by roughly 20 to 50, in essence yielding an overall gain equal to that of a triple cascode. However, it also limits the output voltage swing because the drain voltage of M_1 is equal to $V_{GS,3}$ ($> V_{TH}$), whereas in Figure 7.13(a), it can be as low as $V_{GS,1} - V_{TH}$, an arbitrarily small value. Furthermore, Miller multiplication of $C_{GD,3}$ decreases the magnitude of the pole at the source of M_2, degrading the closed-loop settling behavior of the op amp.

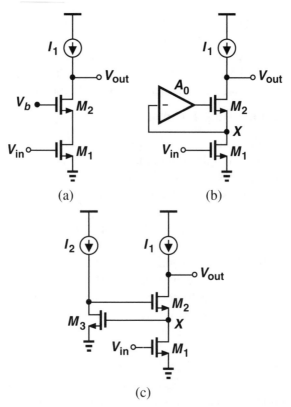

Fig. 7.13   (a) Simple cascode circuit; (b) cascode circuit with gain boosting; (c) simple implementation of (b).

These drawbacks can be alleviated by employing a folded-cascode circuit for A_0 [13]. Shown in Figure 7.14 is such an arrangement, where a folded-cascode amplifier with PMOS input boosts the gain of an NMOS cascode stage. Note that, with proper design, the drain voltage of M_1 can be as low as $V_{GS,1} - V_{TH}$ and the Miller effect at the input of the folded-cascode amplifier is negligible. The overall gain is boosted by more than two orders of magnitude [13].

An important issue in the design of gain-boosted op amps is the effect of the poles of the auxiliary amplifier (A_0 in Figure 7.13) upon the settling behavior of the overall circuit. Bult [13] provides design constraints that ensure fast settling in typical CMOS technologies.

7.1.5 Common-Mode Feedback

In high-gain, fully differential op amps, the output common-mode level must be well-defined even if the circuit is used in closed-loop form. For

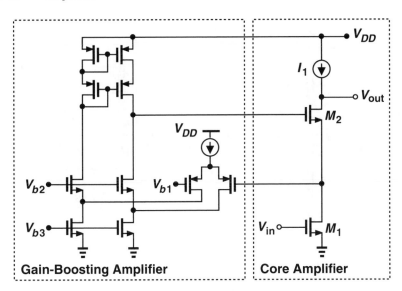

Fig. 7.14   Gain boosting using a folded-cascode amplifier.

example, consider the switched-capacitor circuit in Figure 7.15(a), where during reset S_1 and S_2 are on, providing unity-gain differential feedback. Now suppose the op amp is implemented as a simple differential amplifier, as shown in Figure 7.15(b). In this circuit, when S_1 and S_2 are on, V_X and V_Y are not well-defined because I_{SS} must balance $I_{D3} + I_{D4}$. Thus, for example, if I_{SS} is slightly less than $I_{D3} + I_{D4}$, the output nodes approach V_{DD}, driving M_3 and M_4 into linear region, while the feedback simply senses the *difference* between V_X and V_Y and is therefore unable to correct the CM level. For this reason, differential op amps usually employ a common-mode feedback network (CMFN) to achieve a stable CM level. We discuss a number of approaches to realizing these networks.

The principal issue in the design of CMFNs is that they must have a high *differential* mode rejection; i.e., they must maintain a constant common-mode level even for large differential voltage swings. While it might seem that small variations in the CM level would not lead to any serious problems, in practice common-mode variations may introduce long settling times in the *differential* output. We discuss this point later.

A simple CMFN is shown in the differential amplifier of Figure 7.16(a), where transistors M_5-M_8 provide common-mode feedback. In this circuit, M_5 and M_6 sense the output voltages and produce a CM voltage at their source. This voltage is applied to M_7, setting I_{D8} such that the output CM voltage is equal to $V_{GS5} + V_{GS7}$ (when the differential output is zero).

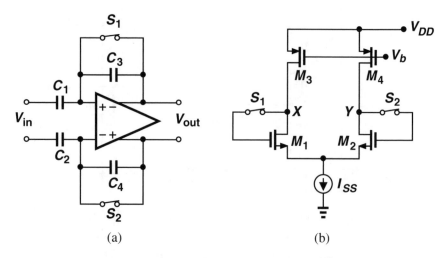

(a)   (b)

Fig. 7.15   (a) Fully-differential switched-capacitor circuit; (b) simple implementation of op amp in (a).

For small differential swings at V_X and V_Y, $I_{D5} + I_{D6}$ is relatively constant and the CM level is fixed. Now suppose the op amp is to produce a large $V_X - V_Y$ (and hence M_1 and M_2 must carry slightly different currents). Since M_5 and M_6 have nonlinear I_D-V_{GS} characteristics, as M_6 begins to turn off, V_P rises, increasing I_{D8} momentarily. The change in I_{D8} is drawn from M_1 and M_2, but *not equally*, because these transistors have slightly different transconductances. As a result, a differential settling component appears at the output solely because the common-mode level has changed. This component can degrade the overall settling behavior if the response of the CMFN is not sufficiently fast.

From the above discussion, we conclude that a CMFN must maintain symmetry even for large differential swings. An example of such a circuit is depicted in Figure 7.16(b), where equal resistors R_1 and R_2 sense V_X and V_Y and reconstruct the CM level at node P [15]. The resulting voltage adjusts the drain currents of M_3 and M_4. In this circuit, since R_1 and R_2 exhibit much more symmetry (i.e., linearity) than M_5 and M_6 in Figure 7.16(a), the CM level remains constant even in the presence of large differential outputs. To eliminate the loading of R_1 and R_2 on the output nodes, these resistors can be preceded by source followers.

Another CMFN used in MOS op amps is shown in Figure 7.16(c) [16], where M_5 and M_6 are biased in the linear region and set the output CM level. Here, the output CM voltage determines the channel resistance of M_5 and M_6 and hence the drain current of M_1 and M_2. The total channel resistance is

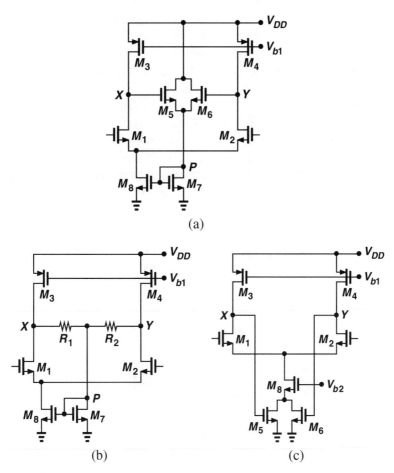

Fig. 7.16 Common-mode feedback using (a) source-coupled pair (b) equal
resistors and (c) MOSFETs in the linear region.

$$R_{\text{tot}} = R_{\text{on},5} || R_{\text{on},6} \tag{7.31}$$

$$= \frac{1}{\mu_n C_{\text{ox}} \dfrac{W}{L}(V_X - V_{\text{TH}})} || \frac{1}{\mu_n C_{\text{ox}} \dfrac{W}{L}(V_Y - V_{\text{TH}})} \tag{7.32}$$

$$= \frac{1}{\mu_n C_{\text{ox}} \dfrac{W}{L}(V_X + V_Y - 2V_{\text{TH}})}, \tag{7.33}$$

indicating that the CM level remains constant if V_X and V_Y change by equal
and opposite amounts.

In switched-capacitor circuits, the CMFN can employ capacitors to sense and correct the output common-mode level [17]. Consider the circuit shown in Figure 7.17(a), where equal capacitors C_1 and C_2 have an initial voltage of V_0 and reproduce a CM level at node P, thus setting the drain current of M_7 such that the output CM level is equal to $V_0 + V_{GS7}$. If the leakage at node P is negligible, the voltage across C_1 and C_2 remains constant, providing symmetric feedback even for large differential outputs. In practice, V_0 must be periodically refreshed by means of a separate circuit, e.g., that of Figure 7.17(b). Here, during refresh, C_1 and C_2 are disconnected from the op amp, charged to a proper voltage, and reconnected to the circuit. In the refresh mode, the op amp output is not valid, but in many data acquisition systems op amps often have some idle time that can be used for this purpose.

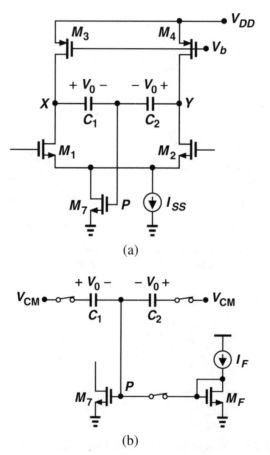

(a)

(b)

Fig. 7.17 (a) Common-mode feedback using capacitors; (b) refresh circuit.

Note that in Figure 7.17(a) M_7 provides only a fraction of the bias current, with the remaining set by I_{SS}. In other words, the CMFN need not control the entire bias current of an op amp. This concept can be applied to all CMFNs and proves useful in optimizing the transient response of the circuit.

7.2 COMPARATORS

The performance of A/D converters that employ parallelism to achieve a high speed strongly depends on that of their constituent comparators. In particular, flash and two-step architectures require great attention to the constraints imposed on the overall system by the large number of comparators. Most such converters utilize voltage comparison, rather than current comparison, because distributing a voltage to a large number of comparators is easier.

Comparison is in effect a binary phenomenon that produces a logic output of ONE or ZERO depending on the polarity of a given input. Figure 7.18(a) depicts the input/output characteristic of an ideal comparator, indicating an abrupt transition (hence infinite gain) at $V_{in,1} - V_{in,2} = 0$. This nonlinear characteristic can be approximated with that of a high-gain amplifier, as shown in Figure 7.18(b). Here, the slope of the characteristic around $V_{in,1} = V_{in,2}$ is equal to the small-signal gain of the amplifier in its active region (A_V), and the output reaches a saturation level if $|V_{in,1} - V_{in,2}|$ is sufficiently large. Thus, the circuit generates well-defined logic outputs if $|V_{in,1} - V_{in,2}| > V_H/A_V$, suggesting that the comparison result is reliable only for input differences greater than V_H/A_V. In other words, the minimum input that can be resolved is approximately equal to V_H/A_V. (The effect of noise is ignored for the moment.) As a consequence, higher resolutions can be obtained only by increasing A_V because V_H, the logical output, cannot be arbitrarily reduced. Since amplifiers usually exhibit strong trade-offs among their speed, gain, and power dissipation, a comparator using a high-gain amplifier will also suffer from the same trade-offs.

Since the amplifiers used in comparators need not be either linear or closed-loop, they can incorporate positive feedback to attain virtually infinite gain. However, to avoid unwanted latch-up, the positive-feedback amplifier must be enabled only at the proper time; i.e., the overall gain of the comparator must change from a relatively small value to a very large value upon assertion of a command.

Figure 7.19 illustrates a typical comparator architecture often utilized in A/D converters. It consists of a preamplifier A_1 and a latch and has two modes of operation: tracking and latching. In the tracking mode, A_1 is enabled to amplify the input difference, hence its output "tracks" the input, while the

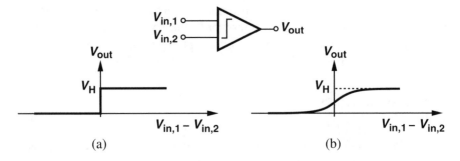

Fig. 7.18   Input/output characteristic of (a) an ideal comparator and (b) a high-gain amplifier.

latch is disabled. In the latching mode, A_1 is disabled and the latch is enabled (strobed) so that the instantaneous output of A_1 is regeneratively amplified and logic levels are produced at V_{out}. Note that it is assumed that the clock edge is sufficiently fast so that the output of A_1 does not diminish during the transition from tracking to latching. Also, if the input to the comparator is constant with time, it is not necessary to disable A_1 in the latching mode. These issues are further discussed below.

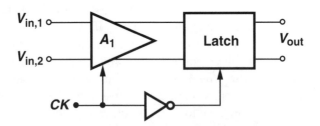

Fig. 7.19   Typical comparator architecture.

Another advantage of the architecture of Figure 7.19 over a simple high-gain amplifier is that the strobe signal (CK) can be used to define a sampling instant at which the polarity of the input difference is stored. As discussed in Chapter 6, this concept is extensively utilized in flash and folding A/D converters to eliminate the need for front-end sample-and-hold circuits.

The use of a latch to perform sampling and amplification of a voltage difference entails an important issue related to the output response in the presence of small inputs: metastability. To explain this issue, we first derive the time response equations of a simple latch.

Figure 7.20(a) shows a latch comprising two identical single-pole inverting amplifiers each with a small-signal gain of $-A_0$ ($A_0 > 0$) and a

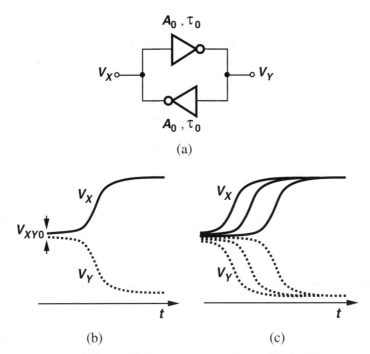

(a)

(b)   (c)

Fig. 7.20   (a) A latch comprising two back-to-back amplifiers; (b) time response of the latch; (c) time response for different values of V_{XY0}.

characteristic time constant of τ_0. For this circuit, we can write:

$$\tau_0 \frac{dV_X}{dt} + V_X = -A_0 V_Y \tag{7.34}$$

$$\tau_0 \frac{dV_Y}{dt} + V_Y = -A_0 V_X. \tag{7.35}$$

Subtracting the second equation from the first and rearranging the terms, we have

$$\tau_0 \frac{d(V_X - V_Y)}{dt} = -(1 - A_0)(V_X - V_Y). \tag{7.36}$$

If the circuit begins with $(V_X - V_Y)|_{t=0} = V_{XY0}$, then

$$V_X - V_Y = V_{XY0} \exp[(A_0 - 1)\frac{t}{\tau_0}]. \tag{7.37}$$

For a typical latch, $A_0 \gg 1$, yielding the important property that the argument of the exponential function is *positive* and hence $V_X - V_Y$ regenerates rapidly.

The regeneration time constant is equal to $\tau_0/(A_0 - 1)$. Figure 7.20(b) shows how the output evolves in time until either of the amplifiers saturates and its gain goes to zero.

An important aspect of latch design is the time needed to produce logic levels after the circuit has sampled a small difference. If $V_X - V_Y$ is to reach a certain value V_{XY1} before it is interpreted as a valid logic level, then the time required for regeneration is

$$T_1 = \frac{\tau_0}{A_0 - 1} \ln \frac{V_{XY1}}{V_{XY0}}. \tag{7.38}$$

Equation (7.38) indicates that T_1 is a function of $\tau_0/(A_0 - 1)$ (and hence the unity-gain bandwidth of each amplifier) as well as the initial voltage difference V_{XY0}. Thus, the circuit has infinite gain if it is given infinite time. In other words, if at the sampling instant V_{XY0} is very small, T_1 will be quite long. This phenomenon is called "metastability" and requires great attention whenever a latch samples a signal that has no timing relationship with the clock. Plotted in Figure 7.20(c) are the V_X and V_Y waveforms with V_{XY0} as a parameter.

Since in most practical cases V_{XY0} is (or can be considered) a random variable, metastability must be quantified in terms of the probability of its occurrence. Suppose in a system using a clock of period $2T_c$, each latch is allowed a regeneration time of T_c. Then, a metastable state occurs if a latch does not produce an output of V_{XY1} within T_c seconds. If the sampled value V_{XY0} has a uniform distribution between $-V_{XY1}$ and $+V_{XY1}$, then the probability of observing a metastable state is [19, 20]

$$P(T_1 > T_c) = \exp \frac{-(A_0 - 1)T_c}{\tau_0}. \tag{7.39}$$

This probability can be lowered by increasing A_0, decreasing τ_0, or pipelining the comparator output.

Before describing various comparator topologies, we define some of their performance metrics.

- Resolution is the minimum input difference that yields a correct digital output. It is limited by the input-referred offset and noise of both the preamplifier and the latch. We call this minimum input 1 LSB (also denoted by V_{LSB}).

- Comparison rate is the maximum clock frequency at which the comparator can recover from a full-scale overdrive and correctly respond to a subsequent 1-LSB input. This rate is limited by the recovery time of the preamplifier as well as the regeneration time constant of the latch.

- Dynamic range is the ratio of the maximum input swing to the minimum resolvable input.

- Kickback noise is the power of the transient noise observed at the comparator input due to switching of the amplifier and the latch.

In addition to these, input capacitance, input bias current, and power dissipation are other important parameters that become critical if a large number of comparators are connected in parallel.

7.2.1 Bipolar Comparators

Figure 7.21 depicts a bipolar implementation of the comparator architecture shown in Figure 7.19. The preamplifier consists of the differential pair Q_1-Q_2 and resistors R_1 and R_2, while the latch comprises Q_3-Q_4 and shares the same resistors. The differential pair and the latch are controlled by CK and $\overline{CK}$ through Q_5 and Q_6, respectively. When CK is high, Q_5 is on and the differential pair tracks the input while Q_6 is off and the latch is disabled. When CK goes low, Q_5 turns off, disabling the input pair, and Q_6 turns on, allowing the latch to establish a positive feedback loop and amplify the difference between V_X and V_Y regeneratively.

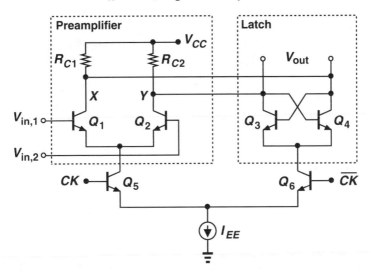

Fig. 7.21   Bipolar implementation of comparator architecture in Figure 7.19.

It is instructive to derive some of the performance metrics of this comparator so as to understand its limitations.

The resolution of the comparator depends on both its input offset voltage and its input-referred noise. The input offset voltage arises from the mismatch between nominally identical devices Q_1-Q_2, R_{C1}-R_{C2}, and Q_3-Q_4. Since mismatch contributions of R_{C1}-R_{C2} and Q_3-Q_4 appear at the output, they are divided by the voltage gain of the differential pair ($g_{m12}R_C$, where R_C is the mean value of R_{C1} and R_{C2}) when referred to the input. For two nominally identical bipolar transistors, the V_{BE} mismatch can be expressed as [3]

$$\Delta V_{BE} = V_T \ln(1 + \frac{\Delta I_S}{I_S}) \tag{7.40}$$

$$= V_T \ln(1 + \frac{\Delta A}{A}) \tag{7.41}$$

$$\approx V_T \frac{\Delta A}{A}, \tag{7.42}$$

where ΔI_S and I_S are the standard deviation and mean value of the saturation current, respectively, and ΔA and A are those of the emitter areas. Equation (7.42) indicates that if, for example, two transistors have a 10% emitter area mismatch, then their V_{BE} mismatch is approximately equal to 2.6 mV at room temperature. Another important observation is that the offset voltage *varies* with temperature; i.e., if it is corrected at one temperature, it may manifest itself at another. We should mention that (7.42) does not include base and emitter resistance mismatch, errors that become increasingly noticeable as devices scale down and are biased at relatively high current densities.

The overall input-referred offset can then be written as

$$V_{OS} = V_T \ln \frac{\Delta A_{12}}{A_{12}} + V_T \frac{\Delta R_C}{R_C} + \frac{1}{g_{m12}R_C} V_T \ln \frac{\Delta A_{34}}{A_{34}}. \tag{7.43}$$

The last term in this equation is negligible if $g_{m12}R_C \gg 1$.

The comparator input-referred noise consists primarily of the thermal and shot noise of Q_1 and Q_2 and the thermal noise of R_{C1} and R_{C2} (neglecting the latch noise). The spectral density of this noise is

$$\frac{\overline{v_n^2}}{\Delta f} = 8kT(r_{b12} + r_{e12} + \frac{1}{2g_{m12}}) + \frac{8kT}{g_{m12}^2 R_C}, \tag{7.44}$$

where r_{b12} and r_{e12} denote base and emitter resistance, respectively, and all the noise components are assumed to be uncorrelated.

Equations (7.43) and (7.44) reveal a number of trade-offs in the design of this comparator. First, to reduce the input offset and r_{e12}, the emitter area of Q_1-Q_2 must increase, thereby increasing the input capacitance. Second, to reduce r_{b12}, the emitter width must increase, again raising the input capacitance. Third, to increase g_{m12}, the bias current must increase, thus increasing the power dissipation. Finally, if R_C is increased, the time constant at nodes X and Y increases and so does the voltage drop across R_{C1} and R_{C2}, thus limiting the input voltage swing. Note that the voltage drop across R_{C1} and R_{C2} should not exceed approximately 300 mV if Q_3 and Q_4 are to remain out of heavy saturation in the latching mode.

To study the comparison rate of this circuit, we first describe the overdrive recovery test, often used as the most stressful assessment of comparator performance. In this test, the input difference toggles between full-scale value V_{FS} and 1 LSB in consecutive clock cycles, yielding the waveforms depicted in Figure 7.22. For a large $\Delta V_{in} = V_{in,1} - V_{in,2}$ (or "overdrive"), the input pair of Figure 7.21 switches completely, steering all of the bias current to one side and producing a large V_{XY}. When ΔV_{in} goes from full-scale to 1 LSB, V_{XY} must "recover" from a large value and become approximately equal to $g_{m12}R_C \times 1$ LSB before the latch is strobed. We note from Figure 7.22 that overdrive recovery has two extreme cases. In the first case, ΔV_{in} goes from $-V_{FS}$ to $+1$ LSB and the output must recover and change polarity. In the second case, ΔV_{in} goes from $-V_{FS}$ to -1 LSB and the output must recover but not change polarity; i.e., it must be free from overshoot. In the first case, if V_{XY} has not changed its polarity before the latch is activated, the latched output will regenerate to its previous value; i.e., the comparator tends to follow residues left from the previous cycle. This phenomenon is called "hysteresis" and results from insufficient time allowed for overdrive recovery.

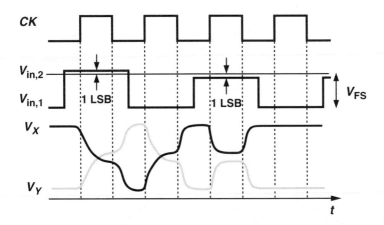

Fig. 7.22   Comparator overdrive test.

From the above discussion, we conclude that, in order for a comparator to respond correctly in an overdrive recovery test, the minimum clock period must allow two phenomena to complete: overdrive recovery in the preamplifier and generation of logic levels after the latch is strobed. In the circuit of Figure 7.21, the preamplifier overdrive recovery can be expressed as

$$V_{XY,ov} = g_{m12}R_C V_{LSB}$$
$$+(V_{CC} - I_{EE}R_C - g_{m12}R_C V_{LSB})\exp\frac{-t}{R_C C_{ov}}, \quad (7.45)$$

where C_{ov} is the average capacitance at nodes X and Y during overdrive recovery (consisting of the collector-base and collector-substrate capacitance of Q_1-Q_4 and the base-emitter junction capacitance of Q_3 and Q_4). The regeneration can be expressed as

$$V_{XY,reg} = V_{XY0}\exp\frac{(g_{m34}R_C - 1)t}{C_{reg}}, \quad (7.46)$$

where V_{XY0} is the difference between V_X and V_Y when regeneration begins and C_{reg} is the average capacitance at nodes X and Y during regeneration (consisting of C_{ov} and the base-emitter diffusion capacitance of Q_3 and Q_4) [19].

The dynamic range of the comparator is given by the ratio of the maximum input swing and V_{LSB} and can be calculated by noting that the input common-mode level $V_{in,CM}$ is limited as follows:

$$2V_{BE} + V_{SEE} \leq V_{in,CM} \leq V_{CC}, \quad (7.47)$$

where V_{SEE} is the minimum voltage required across the current source I_{EE} and it is assumed that $I_{EE}R_C \leq 300$ mV so that Q_1 and Q_2 do not saturate heavily when the input common-mode level reaches V_{CC}.

Another important property of comparators is their kickback noise. Figure 7.23 illustrates how this noise is generated. Suppose the circuit is in the latching mode; i.e., the input pair is off. In the transition to tracking, CK goes high and turns Q_5 on, pulling current from Q_1 and Q_2. However, since Q_1 and Q_2 are initially off, this current first flows through their base-emitter junction, giving rise to a large current spike at $V_{in,1}$ and $V_{in,2}$. The magnitude of this current is approximately equal to half I_{EE} before Q_1 and Q_2 turn on and provide current gain. The duration of this spike depends on the time constant at the input and may extend from one cycle to the next, thereby corrupting the analog input. For example, if $I_{EE} = 200$ μA, in a flash ADC with 256 comparators the kickback noise amplitude may reach tens of milliamperes. This noise can take a long time to decay to below 1 LSB.

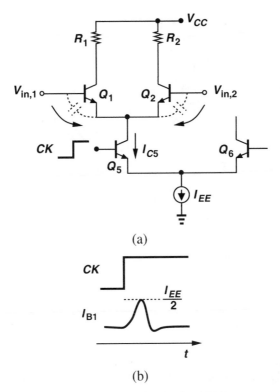

(a)

(b)

Fig. 7.23   Generation of kickback noise in a bipolar comparator.

The comparator of Figure 7.21 exhibits a nonlinear input capacitance as a function of the input difference, as illustrated in Figure 7.24. If $V_{in,1}$ is more negative than $V_{in,2}$ by several V_T, Q_1 is off and the input capacitance is equal to $C_{jc,1} + C_{je,1}$ (for input frequencies much less than f_T of transistors, so that the impedance seen at the emitter of Q_2 is small). As $V_{in,1}$ approaches $V_{in,2}$, Q_1 turns on, introducing a base-emitter diffusion capacitance $C_D = g_m \tau_F$, where τ_F is the base transit time. If $V_{in,1}$ exceeds $V_{in,2}$ by several V_T, Q_2 turns off and Q_1 operates as an emitter follower. In this region, the input capacitance is approximately equal to $C_{jc,1}$ plus a small fraction of $C_{je,1} + C_D$ and increases with $V_{in,1}$ because $C_{jc,1}$ experiences less reverse bias. In a flash A/D converter, for a given input voltage most of the comparators operate in either region 1 or 3, with only a few in region 2. As a result, the converter's input capacitance arises primarily from C_{jc} and C_{je} of the transistors (and interconnect capacitance). The variation of input capacitance with the input voltage causes input-dependent delay and hence harmonic distortion (Chapter 6).

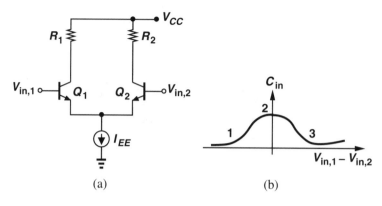

Fig. 7.24   (a) Input stage of a bipolar comparator; (b) small-signal input capacitance versus input differential voltage.

Another important parameter of the comparator of Figure 7.21 is its input bias current. In the tracking mode, this current varies between zero and I_{EE}/β as the input difference changes, and in the latching mode it is zero. As discussed in Chapter 6, in a flash converter, the input bias current of comparators introduces a nonlinear variation in the reference ladder tap voltages.

The limitations described above for the comparator of Figure 7.21 can be significantly relaxed through the use of circuit techniques. In particular, the input differential pair can be preceded with another stage to suppress kickback noise and provide more gain, while the latch can employ emitter followers to enhance the regeneration speed and allow larger voltage swings.

Shown in Figure 7.25 is a comparator circuit often utilized in flash ADCs [18]. It consists of an input stage, a switched differential pair, and a latch comprising Q_9-Q_{12}. The input stage serves the following purposes: (1) it suppresses the kickback noise to acceptably low levels; (2) it provides a relatively high gain, thereby lowering the offset contributed by the latch and

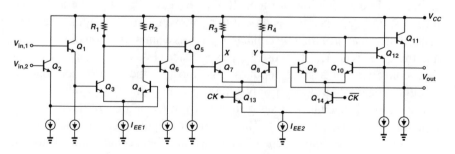

Fig. 7.25   Improved bipolar comparator design.

improving metastability behavior; (3) it exhibits less input capacitance and less feedthrough from one input to the other; (4) its input bias current is relatively constant and can be canceled if necessary. These merits are attained at the cost of larger power dissipation, complexity, and some reduction in small-signal bandwidth. Note that the input offset voltage of the input stage is higher than that of a simple differential pair because the V_{BE} mismatch of Q_1-Q_4 appears directly at the input. By the same token, the input noise is larger as well.

The emitter followers Q_{11} and Q_{12} used in the latch section of Figure 7.25 improve the performance in several ways. First, they reduce the loading effect of the parasitic capacitances of Q_9 and Q_{10} on nodes X and Y, thus enhancing both the small-signal bandwidth and the regeneration speed. Second, they allow larger voltage swings at nodes X and Y because, unlike the circuit of Figure 7.21, the regenerative pair does not enter saturation for swings as large as V_{BE}. Third, they provide a low output impedance for driving the following stage.

In order to increase the input dynamic range, emitter followers Q_1 and Q_2 can be removed from the input stage, and the maximum voltage drop across R_1 and R_2 can be limited to a few hundred millivolts. In this way, the input common-mode level can vary between V_{CC} and $V_{BE} + V_{IEE1}$, yielding a wider input range. The input offset and noise will be less as well. Such a circuit, however, exhibits larger analog input feedthrough and variable input bias current.

Several variants of the comparator circuit shown in Figure 7.25 have been proposed [19, 21, 22]. Particularly interesting is the "high-level clocking" approach used to reduce the kickback noise [19, 21]. In this circuit, illustrated in simplified form in Figure 7.26, rather than turning off the input differential pair, the clock simply steers the current from cascode devices Q_5-Q_6 to the latch Q_3-Q_4. Thus, when CK is high, the circuit is in the tracking mode, and when CK goes low, the circuit enters the latching mode. Note that even though the input pair is never disabled, the latch is not disturbed by the analog input after CK goes low because the current steered to the latch is relatively independent of the analog input voltage. This is important if the clock is to define a sampling instant in an ADC with no front-end SHA.

In this circuit, the input pair does not switch and the kickback noise results only from the transients at nodes A and B due to the switching of Q_5-Q_8. Adding a resistor between A and B decreases the amplitude of these transients and improves the recovery at these nodes. The resistor must be roughly an order of magnitude greater than R_1 and R_2 so that it does not reduce the gain significantly.

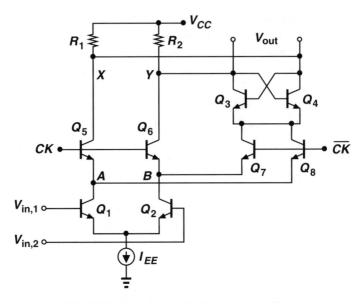

Fig. 7.26 Comparator with high-level clocking.

7.2.2 CMOS Comparators

While the techniques described above substantially improve the performance of bipolar comparators, they do not necessarily yield a high performance if directly applied to the design of CMOS comparators. This is because CMOS devices generally exhibit much smaller transconductance and larger offset than bipolar transistors, thus demanding different design style and circuit topologies.

In order to understand the issues related to CMOS comparator design, let us examine a simple comparator circuit, such as that of Figure 7.27, and compare its performance metrics with those of the bipolar comparators described above. This circuit consists of a differential pair M_1-M_2 and a latch pair M_5-M_6, both sharing the cross-coupled load M_3-M_4. In the tracking mode, CK is low, the input pair is enabled, and M_9 is on, preventing M_3 and M_4 from latch-up. At the same time, M_8 is off, disabling M_5 and M_6. To perform sampling and latching, CK goes high, disabling the input pair, turning off M_9, and enabling M_5 and M_6. Subsequently, M_3-M_6 amplify V_{XY} to rail-to-rail levels.

Transistor M_9 plays two important roles in the circuit. First, it controls the gain in the tracking mode by providing a resistive path between nodes X and Y. The reader can easily show that the gain of the circuit consisting of

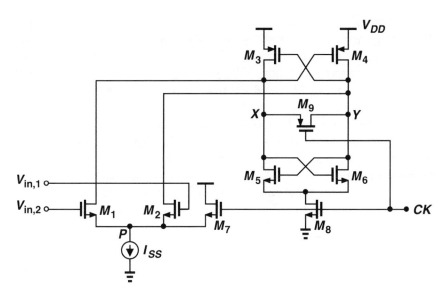

Fig. 7.27   Simple CMOS comparator.

M_1-M_4 and M_9 is equal to

$$A_V = \frac{g_{m12} R_{on,9}}{2 - g_{m34} R_{on,9}}, \tag{7.48}$$

where $R_{on,9}$ is the on-resistance of M_9. As a safe choice, $g_{m34} R_{on,9} \approx 1$, yielding $A_V \approx g_{m12} R_{on,9}$.

The second role of M_9 is to improve the recovery at nodes X and Y when the circuit goes from latching to tracking mode. As M_9 turns on, it experiences a large V_{GS} because either X or Y is at the supply potential. The resulting current pulls X and Y together and rapidly drives V_{XY} to near zero.

We now calculate the performance metrics of this comparator. The input offset voltage arises from mismatches in nominally identical devices M_1-M_2, M_3-M_4, and M_5-M_6. For a simple MOS differential pair, the input offset voltage is

$$V_{OS} = \frac{1}{2}(V_{GS} - V_{TH})(\frac{\Delta W}{W} - \frac{\Delta L}{L}) + \Delta V_{TH}, \tag{7.49}$$

where $\Delta W/W$ and $\Delta L/L$ represent the relative mismatch in the width and length of the devices, respectively [3]. This should be compared with (7.41), where the offset has a *logarithmic* dependence on dimension mismatch and no counterpart for ΔV_{TH}. More importantly, in (7.42) the relative dimension mismatches are multiplied by V_T (≈ 26 mV at room temperature), whereas in (7.49) they are multiplied by $V_{GS} - V_{TH}$ (typically greater than 0.5 V).

In the circuit of Figure 7.27, the offset resulting from the mismatch between M_3 and M_4 is multiplied by g_{m34}/g_{m12} when referred to the input. Similarly, the offset of M_5 and M_6 is multiplied by g_{m56}/g_{m12}, where g_{m56} is the transconductance of M_5 and M_6 at the moment they turn on. Note that when CK goes high and M_8 turns on, M_5 and M_6 experience a $V_{GS} - V_{TH}$ of several volts and hence, from (7.49), exhibit a large offset voltage. Since the preamplifier gain is usually less than 5, the offset contribution of the latch is quite significant.

To calculate the input-referred thermal noise in the tracking mode, note that the noise contribution of M_3 and M_4 at X and Y can be written as

$$\frac{\overline{v_{n34}^2}}{\Delta f} = 4kT \frac{4}{3 g_{m34}} \frac{g_{m34} R_{on,9}}{2 - g_{m34} R_{on,9}}. \tag{7.50}$$

This noise and that due to $R_{on,9}$ are divided by A_v^2 when referred to the input. Thus, the total input thermal noise is

$$\frac{\overline{v_n^2}}{\Delta f} = 4kT \left(\frac{4}{3} \frac{2 - g_{m34} R_{on,9}}{g_{m12}^2 R_{on,9}} + \frac{4}{g_{m12}^2 R_{on,9}} + \frac{4}{3 g_{m12}} \right). \tag{7.51}$$

The above observations reveal a number of trade-offs in the circuit of Figure 7.27. To minimize the noise and offset contribution of M_3-M_6, the transconductance of M_1 and M_2 must increase—at the cost of higher input capacitance or smaller common-mode range—or the transconductance of M_3-M_6 must decrease, slowing down the regeneration at nodes X and Y.

The comparison rate of this circuit is determined by overdrive recovery at X and Y and the regeneration speed of M_3-M_6, both of which are typically slower than those of bipolar comparators. The recovery time constant at nodes X and Y is roughly equal to

$$\tau_{rec} \approx 2 R_{on,9} C_{tot}, \tag{7.52}$$

where C_{tot} represents the total parasitic capacitance seen at each of the nodes X and Y. To decrease τ_{rec}, $R_{on,9}$ must decrease, which in turn reduces A_v.

The regeneration time constant of the latch is

$$\tau_{reg} \approx \frac{C_{tot}}{g_{m34} + g_{m56}}. \tag{7.53}$$

Note that if M_3-M_6 are made wider to increase $g_{m34} + g_{m56}$, then C_{tot} also increases. Thus, τ_{reg} can be considered relatively constant for a given technology.

The input common-mode range of this circuit is given by $V_{DD} - V_{GS,34} + V_{THN} - V_{GS,12} - V_{ISS}$, where V_{ISS} is the minimum voltage required across I_{SS}. It is seen that M_1-M_4 must be sized and biased such that their gate-source voltage does not limit the input range excessively.

The input capacitance of this comparator varies as a function of the input difference in a manner similar to that shown in Figure 7.24(b) for bipolar comparators. However, since a MOS differential pair typically requires more than 1 V of differential input to completely steer the tail current to one side, the comparator input capacitance is relatively constant for most of the input range.

The kickback noise produced at the input of the comparator shown in Figure 7.27 arises from two sources: (1) transients at X and Y when V_X and V_Y are regeneratively amplified to approach the rails, as well as when X and Y are pulled together by M_9; (2) transients at P when this node goes from low to high, and vice versa. The first type of transients couple to the gate of M_1 and M_2 through $C_{GD,1}$ and $C_{GD,2}$, and the second through $C_{GS,1}$ and $C_{GS,2}$.

The large offsets of MOS devices generally limit the resolution of circuits such as that in Figure 7.27 to approximately 6 bits. As a result, CMOS ADCs with resolutions of 8 bits and above usually employ offset cancellation techniques to reduce the minimum resolvable input voltage. This is in contrast with bipolar comparators, which are used with no offset cancellation for resolutions as high as 10 bits.

The design of offset-canceled comparators is described in Chapter 8.

7.2.3 BiCMOS Comparators

The availability of both bipolar and CMOS devices in BiCMOS technologies makes it possible to design comparators with more relaxed speed-power or speed-resolution trade-offs than pure bipolar or CMOS comparators. The speed-power trade-off can be improved through the use of bipolar transistors in the signal path by turning them on only when required, thus saving power [23]. The speed-resolution trade-off can be eased by applying offset cancellation to bipolar transistors and hence achieve smaller offsets than that attained in either bipolar or CMOS comparators [25].

In order to reduce the power dissipation, the design of comparators can be based on the concept of "charge steering." Originally used in CMOS memory sense amplifiers [24], this concept has evolved from current-steering circuits by replacing resistors with capacitors and currents with charge. We describe this evolution by first considering the bipolar comparator of Figure 7.21, repeated in Figure 7.28(a). In this circuit, the mode of operation is controlled by steering the tail current I_{EE} between the input pair Q_1-Q_2 and the latch Q_3-Q_4. The tail current is constant, thus giving a continuous power dissipation of $I_{EE}V_{CC}$, and the voltage at output nodes X and Y is established by the flow of I_{EE} through R_1 or R_2.

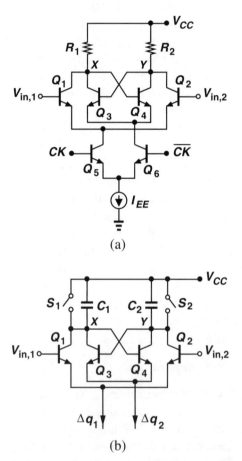

(a)

(b)

Fig. 7.28   (a) Current-steering comparator; (b) charge-steering comparator.

Now consider the circuit shown in Figure 7.28(b), where C_1 and C_2 function as load devices and the pairs Q_1-Q_2 and Q_3-Q_4 are activated by charge packets Δq_1 and Δq_2. This circuit has three modes of operation: precharge, input sampling, and regeneration. A typical comparison proceeds as follows. First, nodes X and Y are precharged to V_{CC}. Next, a charge packet, Δq_1, is drawn from Q_1 and Q_2, biasing these transistors for a short period of time and producing an amplified version of the input difference across C_1 and C_2. Subsequently, a charge packet, Δq_2, is pulled from Q_3 and Q_4, activating the latch momentarily and amplifying V_{XY} regeneratively.

The charge-steering comparator can be viewed as a "discrete-time" version of the current-steering configuration; i.e., it operates with short pulses of current rather than a continuous bias. As a consequence, it can dissipate less power even at high frequencies.

The dynamic power dissipation of the circuit is given by the value of Δq_1 and Δq_2 and the comparison rate. The minimum value of these charge packets is determined by the required gain and output voltage swings, which themselves depend on C_1 and C_2. These capacitors must be large enough to suppress the charge injection mismatch of S_1 and S_2.

In order to better understand the behavior of this comparator, we perform a simple calculation to estimate the small-signal gain of the input differential pair. Consider the circuit shown in Figure 7.29(a), where M_1, M_2, and C_P constitute a "charge pump," pulling charge from Q_1 and Q_2 when CK goes low. Let us assume that the input common-mode level remains close to V_{CC} and the on-resistance of M_1, $R_{on,1}$, is relatively constant. Using the equivalent circuit shown in Figure 7.29(b), we can write

$$I_P(t) \approx \frac{V_{CC} - V_{BE}}{R_{on,1}} \exp \frac{-t}{R_{on,1} C_P}, \tag{7.54}$$

where variations in V_{BE} are neglected with respect to V_{CC}. Since the bias current of Q_1 and Q_2 varies during the amplification mode, the transconductance of these devices must be expressed as a function of time:

$$g_m(t) = \frac{I_P(t)}{2V_T} \tag{7.55}$$

$$= \frac{V_{CC} - V_{BE}}{2R_{on,1} V_T} \exp \frac{-t}{R_{on,1} C_P}, \tag{7.56}$$

where $V_{in,1} - V_{in,2}$ is assumed to be small so that $I_{C1} \approx I_{C2}$ at all time. The final differential output voltage is therefore equal to

$$V_{XY} = \int_0^\infty \frac{g_m(t)(V_{in,1} - V_{in,2})}{C} \, dt, \tag{7.57}$$

where $C = C_1 = C_2$. If $V_{in,1} - V_{in,2}$ is constant, then it follows form (7.56) and (7.57) that the small-signal gain is

$$\frac{V_{XY\infty}}{V_{in,1} - V_{in,2}} = \frac{V_{CC} - V_{BE}}{2V_T} \frac{C_P}{C}, \tag{7.58}$$

where $V_{XY\infty}$ is the final value of V_{XY}. For example, if $V_{CC} - V_{BE} \approx 4$ V, $V_T = 26$ mV, and $C_P = 0.5\, C$, then the gain is roughly equal to 38.

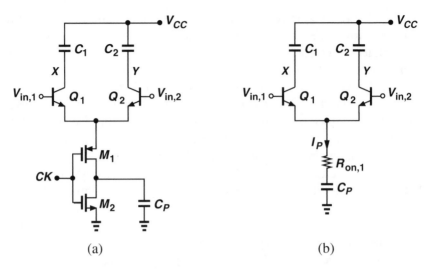

Fig. 7.29   (a) Simplified charge-steering amplifier; (b) equivalent circuit of (a).

While a power efficient configuration, the circuit of Figure 7.28(b) is typically quite slower than that in Figure 7.28(a). Several reasons account for this difference. First, the delays required between the clock edges that turn off S_1 and S_2 and produce Δq_1 and Δq_2 impose an upper bound on the conversion rate. Second, C_1 and C_2 slow down the regeneration at nodes X and Y. Third, the clocks controlling S_1, S_2, and the charge packets typically need rail-to-rail swings and hence suffer from longer transition times than CK and $\overline{CK}$ in Figure 7.28(a), which usually require only a few hundred millivolts of swing.

For the simple charge pump circuit shown in Figure 7.29, the amount of charge drawn from the differential pair depends on the input common-mode level, making the gain and output voltage swings have the same dependence. If the variation in the common-mode level is comparable with $V_{CC} - V_{BE}$, then the charge pump circuit must be modified so that it operates as a constant current source during pumping.

The precharge and sampling modes in a charge-steering comparator can be converted to a single tracking mode if the input differential pair is replaced with MOS switches, as depicted in Figure 7.30(a). Here, the input is sampled on C_1 and C_2 and subsequently amplified by the latch. However, if $|V_{in,1} - V_{in,2}|$ exceeds roughly 0.7 V, the base-collector junction of Q_1 or Q_2 is forward-biased, drawing current from the inputs. To avoid this problem, the base of each transistor can be capacitively coupled to the collector of the other, as depicted in Figure 7.30(b) [23]. In this circuit, first the input difference is sampled on C_3 and C_4, then the collectors of Q_1 and Q_2 are released from V_{CC}, and finally the latch is activated by a charge packet drawn from Q_1 and Q_2. Note that the loop gain of the latch is attenuated by the voltage division due to C_4 and the input capacitance of Q_1 (and C_3 and the input capacitance of Q_2).

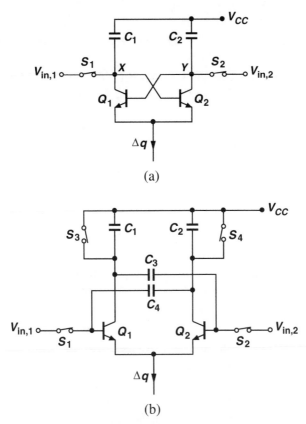

(a)

(b)

Fig. 7.30   Charge-steering comparator without precharge cycle. (b) Modified version of (a) allowing large $V_{in,1} - V_{in,2}$.

REFERENCES

[1] R. Jewett, J. Corcoran, and G. Steinbach, "A 12b 20 MS/s Ripple-Through ADC," *ISSCC Dig. Tech. Pap.*, pp. 34-35, Feb. 1992.

[2] P. Petschacher et al., "A 10-b 75-MSPS Subranging A/D Converter with Integrated Sample and Hold," *IEEE J. Solid-State Circuits*, vol. SC-25, pp. 1339-1346, Dec. 1990.

[3] P. R. Gray and R. G. Meyer, *Analysis and Design of Analog Integrated Circuits*, Third Edition, John Wiley and Sons, New York, 1992.

[4] P. Quinn, "A Cascode Amplitude Nonlinearity Correction Technique," *ISSCC Dig. Tech. Pap.*, pp. 188-189, Feb. 1981.

[5] S. Simpkins and W. Gross, "Cascomp Feedforward Error Correction in High Speed Amplifier Design," *IEEE J. Solid-State Circuits*, vol. SC-18, pp. 762-764, Dec. 1983.

[6] S. M. Sze, *Physics of Semiconductor Devices*, Second Edition, John Wiley and Sons, New York, 1981.

[7] M. Nayebi and B. A. Wooley, "A 10-Bit Video BiCMOS Track-and-Hold Amplifier," *IEEE J. Solid-State Circuits*, vol. SC-24, pp. 1507-1516, Dec. 1989.

[8] J. E. Solomon, "The Monolithic Op Amp: A Tutorial Study," *IEEE J. Solid-State Circuits*, vol. SC-9, pp. 314-332, Dec. 1974.

[9] P. R. Gray and R. G. Meyer, "MOS Operational Amplifier Design—A Tutorial Overview," *IEEE J. Solid-State Circuits*, vol. SC-17, pp. 969-982, Dec. 1982.

[10] B. K. Ahuja, "An Improved Frequency Compensation Technique for CMOS Operational Amplifiers," *IEEE J. Solid-State Circuits*, vol. SC-18, pp. 629-633, Dec. 1983.

[11] J. N. Babanezhad, "A Rail-to-Rail CMOS Op Amp," *IEEE J. Solid-State Circuits*, vol. SC-23, pp. 1414-1417, Dec. 1988.

[12] S. H. Lewis et al., "A Pipelined 9-Stage Video-Rate Analog-to-Digital Converter," *Proc. CICC*, pp. 26.4.1-26.4.4, May 1991.

[13] K. Bult and G. J. G. M. Geelen, "A Fast-Settling CMOS Op Amp for SC Circuits with 90-dB DC Gain," *IEEE J. Solid-State Circuits*, vol. SC-25, pp. 1379-1384, Dec. 1990.

[14] E. Sackinger and W. Guggenbuhl, "A High-Swing, High-Impedance MOS Cascode Circuit," *IEEE J. Solid-State Circuits*, vol. SC-25, pp. 289-298, Feb. 1990.

[15] M. Banu, J. A. Khoury, and Y. Tsividis, "Fully Differential Operational Amplifiers with Accurate Output Balancing," *IEEE J. Solid-State Circuits*, vol. SC-23, pp. 1410-1414, Dec. 1988.

[16] T. C. Choi et al., "High-Frequency CMOS Switched-Capacitor Filters for Communications Applications," *IEEE J. Solid-State Circuits*, vol. SC-18, pp. 652-664, Dec. 1983.

[17] D. Sendrowicz et al., "A Family of Differential NMOS Analog Circuits for a PCM Codec Filter Chip," *IEEE J. Solid-State Circuits*, vol. SC-17, pp. 1014-1023, Dec. 1982.

[18] Y. Akazawa et al., "A 400 MSPS 8b Flash AD Conversion LSI," *ISSCC Dig. Tech. Pap.*, pp. 98-99, Feb. 1987.

[19] B. Zojer, R. Petschacher, and W. A. Luschnig, "A 6-Bit 200-MHz Full Nyquist A/D Converter," *IEEE J. Solid-State Circuits*, vol. SC-20, pp. 780-786, June 1985.

[20] H. J. M. Veendrick, "The Behavior of Flip-Flops Used as Synchronizers and Prediction of Their Failure Rate," *IEEE J. Solid-State Circuits*, vol. SC-15, pp. 169-176, April 1980.

[21] R. J. van de Plassche and P. Baltus, "An 8-Bit 100-MHz Full-Nyquist Analog-to-Digital Converter," *IEEE J. Solid-State Circuits*, vol. SC-23, pp. 1334-1344, Dec. 1988.

[22] V. E. Garuts et al., "A Dual 4-Bit 1.5 GS/s Analog-to-Digital Converter," *Proc. BCTM*, pp. 141-144, Sept. 1988.

[23] P. J. Lim and B. A. Wooley, "An 8-bit 200-MHz BiCMOS Comparator," *IEEE J. Solid-State Circuits*, vol. SC-25, pp. 192-199, Feb. 1990.

[24] K. U. Stein, A. Sihling, and E. Doering, "Storage Array and Sense/Refresh Circuit for Single-Transistor Memory Cells," *IEEE J. Solid-State Circuits*, vol. SC-7, pp. 336-340, Oct. 1972.

[25] B. Razavi and B. A. Wooley, "Design Techniques for High-Speed High-Resolution Comparators," *IEEE J. Solid-State Circuits*, vol. SC-27, pp. 1916-1926, Dec. 1992.

8

Precision Techniques

As data conversion interfaces are designed for higher precisions, the nonidealities that accompany monolithic devices become more problematic. Effects such as mismatch, nonlinearity, and finite intrinsic gain limit the "raw" resolution to approximately 6 bits in CMOS technology and 10 bits in bipolar technology. For higher resolutions, it is often necessary to correct for these effects by means of circuit or algorithmic techniques. Such techniques are applied to individual building blocks to improve their precision, as well as to the overall architecture to make its input/output characteristic approach the ideal. In addition to the regular mode of operation, circuits employing some of these techniques typically require a dedicated period to carry out cancellation or calibration, thereby complicating the system's timing scheme.

In this chapter, we describe a number of methods often used to enhance the precision or relax the speed-precision trade-offs of data acquisition systems. These include comparator and op amp offset cancellation, DAC and ADC calibration, and range overlap with digital correction.

8.1 COMPARATOR OFFSET CANCELLATION

The resolution limitations discussed in Chapter 7 for comparators mandate offset cancellation in high-precision systems. This is particularly crucial in pure CMOS circuits because of the large mismatches of MOS devices and is also important in bipolar and BiCMOS circuits if resolutions above 10 bits are required.

Traditionally, laser wafer trimming and fuse techniques have been used in bipolar circuits to cancel offsets [1]. However, these techniques suffer from several drawbacks: substantial area and cost penalty, failure to maintain the cancellation over temperature and time, and sensitivity to package stress. Furthermore, these methods cannot be easily applied to CMOS circuits because the large offsets of MOS transistors require a much wider correction range than do their bipolar counterparts.

The need for reliable offset cancellation has led to autozeroing techniques in CMOS and BiCMOS comparators, wherein the offset is periodically sensed, stored, and added to the input in such a way as to cancel itself. Of the variety of comparator offset cancellation techniques [2, 3, 4], we discuss four here. The general architecture using any of these techniques is similar to that of Figure 7.19, consisting of a preamplifier and a latch. Each circuit has three modes of operation: offset cancellation, tracking, and latching. The challenge is to minimize the input offset contributed by the preamp and the latch without compromising other aspects of the performance.

8.1.1 Input Offset Storage

This technique (denoted by IOS) measures the input offset of the preamplifier by closing a unity-gain feedback loop around it and stores the resulting offset on capacitors in series with the input. Figure 8.1 depicts a configuration employing this technique. The circuit operates as follows. During offset cancellation, S_1-S_4 are on, S_5 and S_6 are off, nodes A and B are grounded, a unity-gain feedback loop is established around A_0, and the input offset is stored on C_1 and C_2. During tracking, S_1-S_4 are off, S_5 and S_6 are on, the feedback loop is open, and the preamplifier senses the analog input and amplifies the difference. In the latching mode, the latch is strobed so as to regeneratively amplify the difference produced at the preamplifier output, hence providing logic levels at V_{out}.

The residual offset (the offset after cancellation) of this topology can be calculated as follows. For the feedback circuit in the offset cancellation mode, we have

$$(V_{PQ} - V_{\text{OSA}})(-A_0) = V_{XY} = V_{PQ}, \tag{8.1}$$

where V_{OSA} is the input offset of the preamplifier. Thus,

$$V_{XY} = \frac{A_0}{1 + A_0} V_{\text{OSA}}. \tag{8.2}$$

This voltage is stored on C_1 and C_2. When the feedback loop opens, the preamplifier output remains the same if S_3 and S_4 exhibit perfect matching.

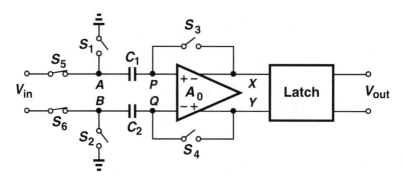

Fig. 8.1 Input offset storage.

Therefore, for a zero difference at the comparator input, the preamplifier output is approximately equal to V_{OSA}. In contrast, if there were no offset cancellation, the preamplifier output would be equal to $A_0 V_{OSA}$. Thus, IOS reduces the effect of the preamplifier offset by approximately a factor of A_0.

In addition to V_{OSA}, the latch offset, V_{OSL}, and charge injection mismatch between S_3 and S_4, Δq, contribute to the comparator input offset. Of these two, V_{OSL} is divided by A_0 when referred to the input, whereas Δq appears directly at the input, yielding a total offset of

$$V_{OS(tot)} = \frac{V_{OSA}}{1 + A_0} + \frac{\Delta q}{C} + \frac{V_{OSL}}{A_0}, \qquad (8.3)$$

where C is the value of C_1 and C_2 (assumed equal here).

The IOS topology has two important features. First, it allows a rail-to-rail common-mode level at the comparator input (when using complementary MOS switches for S_5 and S_6), thereby achieving a wide dynamic range. Second, it performs the cancellation in a closed loop, thus driving the preamplifier to its active region as well as augmenting its overdrive recovery.

Despite these features, IOS suffers from several drawbacks. The offset contributed by the preamplifier can be reduced only by employing a high gain, degrading the power-speed trade-off. Furthermore, since the value of input coupling capacitors is dictated by charge injection mismatch, kT/C noise, and attenuation considerations, C_1 and C_2 (and hence their parasitics) are usually quite large. Note that a large input capacitance not only slows down the preceding circuit (e.g., the front-end sample-and-hold amplifier) but also introduces a substantial amount of noise at the input when the circuit enters the tracking mode and nodes A and B must charge to proper voltages. This issue is particularly important if many comparators are connected to a resistor ladder.

8.1.2 Output Offset Storage

This technique (denoted by OOS) measures the *output-referred* offset of the preamplifier by grounding its inputs and stores the result on capacitors in series with the preamplifier output. Illustrated in Figure 8.2, OOS operates as follows. During offset cancellation, nodes A, B, X, and Y are grounded and the preamplifier offset is amplified and stored on C_1 and C_2. In the tracking mode, S_1-S_4 turn off and S_5 and S_6 turn on. The circuit thus senses and amplifies the input difference, generating a differential voltage at the input of the latch. Next, in the latching mode, the latch is strobed to amplify its input voltage and produce logic levels at V_{out}.

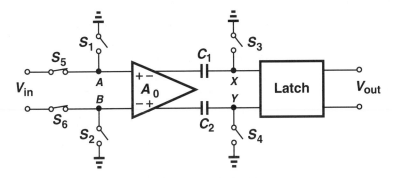

Fig. 8.2   Output offset storage.

To calculate the residual offset of OOS, first note that in the cancellation mode, the latch input voltage is zero; i.e., a zero difference at the comparator input gives a zero difference at the latch input. Consequently, the offset of the preamplifier is completely canceled, a feature in contrast with IOS. Thus, the total offset results from only the charge injection mismatch between S_3 and S_4 and the latch offset:

$$V_{OS(tot)} = \frac{\Delta q}{A_0 C} + \frac{V_{OSL}}{A_0}. \tag{8.4}$$

Note that the effect of charge injection mismatch is divided by A_0, another advantage of OOS over IOS.

In addition to a potentially lower offset, OOS usually exhibits less input capacitance than IOS and is therefore more suitable for parallel ADC architectures. However, it suffers from two drawbacks that do not exist in IOS. Since the preamplifiers used in OOS are open loop, they cannot employ high gain. This is because, in the cancellation mode, the preamplifier may saturate if the product of its input offset and its gain exceeds the maximum voltage swing allowed at its output. As a consequence, OOS typically incorporates a single-stage amplifier with a gain less than 20.

The other drawback of OOS stems from dc coupling at its input and hence limited input common-mode range. This is in contrast with IOS, where rail-to-rail inputs can be accommodated.

8.1.3 Multistage Offset Storage

From the discussion in the previous section it follows that for high resolutions a single stage of OOS is inadequate, while a single stage of IOS with high gain suffers from a long delay. These considerations have led to the use of multistage cancellation techniques in high-resolution comparators [2, 4].

Figure 8.3 illustrates a typical multistage cancellation scheme. The circuit comprises a cascade of capacitively coupled amplifiers followed by a latch. In this example, all amplifiers use OOS, but IOS or a combination of both are also possible. Since the equivalent gain of the overall amplifier is the *product* of the gains of individual stages, a high gain with fast response can be achieved. The circuit operates as follows. In the offset cancellation mode, the inputs of $A_1, \ldots, A_n$ and the latch are grounded, and the offset of each amplifier is stored on capacitors in series with its outputs. In the tracking mode, the input difference is sensed and amplified by $A_1, \ldots, A_n$. In the latching mode, the latch is strobed to produce logic levels at V_{out}.

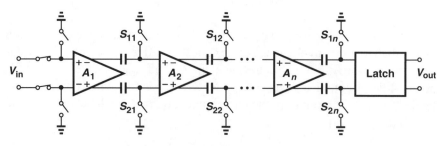

Fig. 8.3   Multistage offset cancellation.

In the circuit of Figure 8.3, the input offset resulting from the latch is equal to

$$V_{\text{OS}} = \frac{V_{\text{OSL}}}{A_1 A_2 \ldots A_n}. \tag{8.5}$$

The number of stages used in this approach depends on the overall gain required to suppress V_{OS} below a certain value. For a given gain, n can be calculated so as to achieve the maximum bandwidth [4], but in practice it is between 2 and 4.

An important source of error was neglected in the above discussion: channel charge injection mismatch of switches S_{11}-S_{2n}. Depending on their exact turn-off timing, these switches may contribute to the input offset in the transition from cancellation to tracking. For example, if S_{11} and S_{21} turn off slightly later than others, their channel charge mismatch introduces an uncanceled offset at the input of A_2. To minimize this type of error, the comparator can utilize sequential clocking [2], wherein the gain stages leave the offset cancellation mode sequentially, A_1 first and A_n last.

Figure 8.4 illustrates this timing arrangement. When CK_1 goes low, A_1 leaves the cancellation mode, whereas $A_2, \ldots, A_n$ are still in that mode. Consequently, the offset due to charge injection mismatch of S_{11} and S_{21} is amplified by A_2 and stored on the capacitors at its output. In other words, during this interval A_2 acts as an OOS stage and hence reduces its residual offset to zero. Next, CK_2 goes low while $CK_3, \ldots, CK_n$ are still high, allowing the offset due to charge injection mismatch of S_{12} and S_{22} to be canceled by the following stage. This sequence continues until CK_n goes low. Thus, the error due to charge injection mismatch of all the switches except S_{1n} and S_{2n} is canceled.

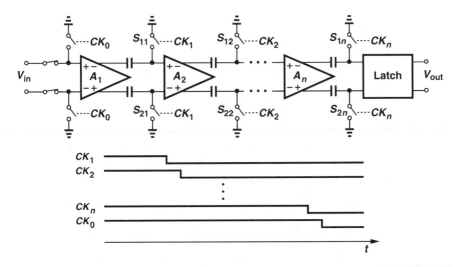

Fig. 8.4   Sequential clocking in multistage comparators.

In designing a multistage comparator, two issues should be considered. First, the delay between the edges of $CK_1, \ldots, CK_n$ should be long enough to allow complete offset storage on the following capacitors. Figure 8.5 depicts typical resistive paths for the offset storage capacitors. With OOS, the total resistance in series with C_{1j} is $R_{oj} + R_{on,j}$, where R_{oj} is the *open-loop* output resistance of A_j and $R_{on,j}$ is the on-resistance of switch S_{1j}. With IOS, the total resistance is $R_{oj} + R_{on,j}/A_{j+1} + R_{o(j+1)}$, where $R_{o(j+1)}$ is the *closed-loop* output impedance of A_{j+1} when configured as a unity-gain feedback amplifier. The resulting time constant in each case imposes a lower bound on the delay between CK_j and CK_{j+1}.

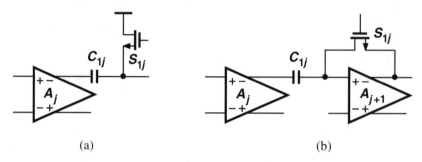

(a)   (b)

Fig. 8.5   Resistive paths associated with offset storage capacitors in a multistage comparator. (a) OOS; (b) IOS.

The second issue relates to the error voltage introduced by charge injection and clock feedthrough when each stage leaves the offset cancellation mode. If the product of this voltage and the gain of a stage is excessively large, the amplifier in that stage is driven out of its active region, hence exhibiting a much lower gain.

Multistage comparators have been implemented in several forms [2, 4, 5]. A simple approach is illustrated in Figure 8.6, where CMOS inverters operate as amplifiers, with IOS applied to all the stages [5]. The simplicity of this configuration has made it quite popular in the resolution range of 8 to 10 bits [6, 7]. An important feature of this topology is its scalability with lower supply voltages, essentially the same as digital CMOS circuits. For example, a 10-bit ADC [7] utilizes such stages extensively to allow operation from 2.5 V with a total power dissipation of 30 mW at 20 MHz.

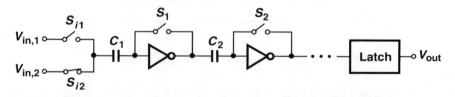

Fig. 8.6   Multistage offset cancellation using CMOS inverters.

This approach nonetheless suffers from several drawbacks. The single-ended operation raises serious concerns regarding charge injection error and supply rejection ratio. In order to ensure that the inverters remain in their high-gain region after the feedback switches turn off, the coupling capacitors must be quite large. Furthermore, the output voltage of each inverter is very sensitive to supply noise because the small-signal input/output gain of inverters is on the same order as the gain from the supply line to their output.

In addition, the power dissipation of the circuit is strongly process- and supply-dependent because it is given primarily by the bias current of inverters when their feedback switches are on, i.e., the current drawn from the supply by two diode-connected MOSFETs in series. Also, as mentioned for IOS, the large capacitance and switching noise seen at the input of this topology give rise to long settling times and are particularly detrimental in flash architectures.

Multistage comparators can also be implemented in fully differential form through the use of differential amplifiers. In contrast with the topology of Figure 8.6, differential circuits are much less sensitive to common-mode effects such as charge injection and supply variation.

Figure 8.7(a) shows a simple CMOS differential pair used in differential comparators. For square-law devices, the gain of this circuit has a square root dependence on device size and is usually limited to less than 10 for reasonable device dimensions. Thus, it is well-suited to OOS. To achieve a higher gain, the load circuit can be modified as shown in Figure 8.7(b), where current sources I_1 and I_2 provide a major portion of the current drawn by M_1 and M_2 [4]. Since the bias current of M_3 and M_4 is less than that of M_1 and M_2, the gain can be made higher than that in Figure 8.7(a). In practice, I_1 and I_2 track I_{SS} within a few percent. As a safe choice, $I_1 = I_2 = 0.75(0.5I_{SS})$, so that M_3 and M_4 do not "starve" in the presence of tracking and mismatch errors.

Another approach to increasing the gain is to incorporate positive feedback with controlled loop gain [8]. Illustrated in Figure 8.7(c), this technique configures the load devices such that they boost the gain without entering latch-up. The overall gain of this circuit is equal to

$$A_v = \frac{g_{m12}}{g_{m34}} \frac{1}{1 - g_{m56}/g_{m34}}, \tag{8.6}$$

where the second fraction represents the gain resulting from positive feedback. Since g_{m56}/g_{m34} is defined only by the ratio of PMOS device dimensions, it can be controlled quite tightly. For example, with $W_{56}/W_{34} = 3/4$ and $L_{56} = L_{34}$, we have $I_{D56}/I_{D34} = 3/4$, and positive feedback boosts the gain by a factor of 4.

While multistage comparator topologies achieve a small input offset, they also pose several problems. First, they suffer from strong trade-off among gain, bandwidth, and power dissipation, often trading the latter two for a

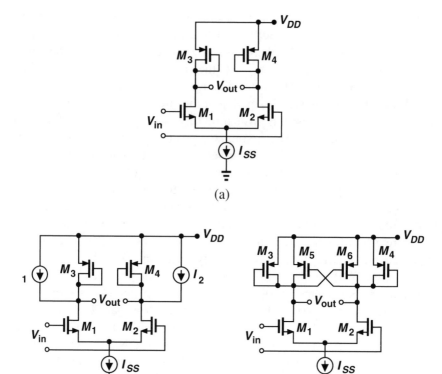

Fig. 8.7 (a) CMOS differential pair; (b) gain boosting using current sources; (c) gain boosting using positive feedback.

high gain to ensure adequate resolution. This is especially acute if the latch exhibits large offsets. Second, the complexity of these comparators and their clocking sequence make them difficult to use in flash or two-step ADCs. Third, the sequential clocking required to suppress the effect of charge injection mismatch both imposes a lower bound on the clock period and faces timing skew issues when clocks must be distributed on a large chip.

8.1.4 Comparators Using Offset-Canceled Latches

The problems described for multistage comparators can be substantially alleviated if the offset of the latch is reduced in a reliable way. With a lower latch offset, the preamplifier need not have a high gain and can therefore be optimized for speed and power dissipation. In this section, we describe a comparator architecture that applies offset cancellation to both the preamplifier and the latch.

Figure 8.8 depicts a simplified block diagram of the comparator [3]. It consists of transconductance amplifiers G_{m1} and G_{m2}, load resistors R_{L1} and R_{L2}, and capacitors C_1 and C_2 placed in a positive feedback loop around G_{m2}. In the offset-cancellation mode, the inputs of G_{m1} and G_{m2} are grounded and their offsets are amplified and stored on C_1 and C_2. In the comparison mode, the inputs are released from ground and the input voltage is sensed. This voltage is amplified by G_{m1} to establish an imbalance at the output nodes and hence at G_{m2} inputs, initiating a fast regeneration around G_{m2}.

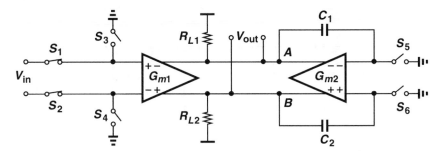

Fig. 8.8   CMOS comparator with offset-canceled latch.

The calibration of this circuit can be viewed as OOS applied to both G_{m1} and G_{m2}, resulting in complete cancelation of their offsets. The important difference is that this topology utilizes the offset-canceled amplifier G_{m2} for regeneration, whereas OOS incorporates an explicit latch that can suffer from a large input offset. Thus, neglecting second-order effects such as charge injection mismatch of S_5 and S_6, this configuration accomplishes zero residual offset while retaining all of the advantages of OOS.

Due to several complications, the configuration of Figure 8.8 is not quite practical if implemented as such. First, the feedback capacitors and their parasitics load the output nodes, reducing the speed. Second, because of the finite on-resistance of S_5 and S_6, the positive feedback loop around G_{m2} is not completely broken in the calibration mode, making the circuit prone to oscillation. More importantly, when S_5 and S_6 turn off to end the calibration, any mismatch in their charge feedthrough can trigger a false regeneration around G_{m2}. Since the feedback is designed for a fast response, this regeneration cannot be overridden by small voltages at the input, hence causing a large input-referred offset. Figure 8.9 illustrates a modified block diagram of the comparator that circumvents these problems. In this circuit, buffers B_1 and B_2 isolate nodes A and B from the feedback capacitors, while switches S_7-S_{10} disable the feedback loop when required. Regeneration then begins only after the input voltage has been sensed and amplified.

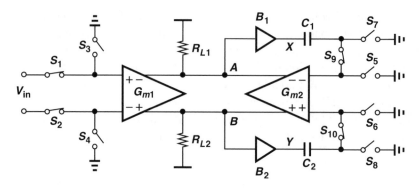

Fig. 8.9   Modified version of the circuit in Figure 8.8.

Since this configuration calibrates both the preamplifier and the latch, its residual offset is primarily caused by charge feedthrough mismatch of switches S_5-S_{10}. The resulting errors are discussed in [3].

8.2   OP AMP OFFSET CANCELLATION

Pipelined A/D converters often require precision op amps to perform inter-stage sampling, subtraction, and amplification [9]. The input-referred offset of these op amps gives rise to differential nonlinearity and must remain well below 1 LSB, usually mandating offset cancellation in high-resolution systems.

The major concern in cancelling the offset of an op amp, unlike that of an open-loop circuit such as a comparator, is the phase margin degradation caused by the cancellation circuit, in particular the offset storage capacitors. While some of the offset cancellation techniques described for comparators in previous sections may be applied to op amps [10], they substantially degrade the closed-loop settling behavior by introducing capacitors in the signal path, i.e., by adding more nondominant poles to the circuit. Since the minimum size of these capacitors (and their parasitics) is dictated by charge injection mismatch and kT/C noise, the magnitude of these poles may not be sufficiently large, thus increasing the settling time markedly.

A more efficient approach is to use an auxiliary amplifier that isolates the signal path from offset storage capacitors while canceling the offset of the main amplifier [11].

A possible implementation of this concept is illustrated in Figure 8.10, where a transconductance amplifier G_{m1} and a transresistance amplifier R constitute the main amplifier and another transconductance amplifier G_{m2} performs offset cancellation. The circuit operates as follows. During offset cancellation, S_1-S_4 are on while S_5 and S_6 are off, inputs of G_{m1} are grounded, and the $G_{m2}R$ loop is closed. The op amp offset is thus stored on C_1 and C_2. During amplification, only S_5 and S_6 are on and G_{m2} simply adds a dc component at the output so as to cancel the op amp offset.

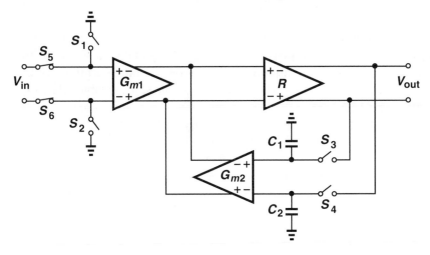

Fig. 8.10   Op amp offset cancellation using an auxiliary amplifier.

To calculate the residual offset of the circuit, we first consider a hypothetical case where the inputs of both G_{m1} and G_{m2} are grounded and the $G_{m2}R$ loop is *open*. In that case, the input offsets of G_{m1} and G_{m2}, denoted by $V_{OS,1}$ and $V_{OS,2}$, respectively, would be amplified by their respective gains and appear at the output as $G_{m1}RV_{OS,1} + G_{m2}RV_{OS,2}$. Now, if the $G_{m2}R$ loop is closed, negative feedback reduces V_{out} by a factor of approximately $G_{m2}R$:

$$V_{out} = \frac{G_{m1}RV_{OS,1} + G_{m2}RV_{OS,2}}{G_{m2}R}. \tag{8.7}$$

This is the overall offset voltage of the circuit *referred to the output* (and is stored on C_1 and C_2). To find the offset referred to the input of G_{m1}, (8.7) should be divided by the gain of the main amplifier $G_{m1}R$:

$$V_{OS} = \frac{V_{OS,1}}{G_{m2}R} + \frac{V_{OS,2}}{G_{m1}R}. \tag{8.8}$$

At the end of offset cancellation, when S_3 and S_4 turn off, their charge injection mismatch results in a differential error voltage, ΔV, on C_1 and C_2. This error appears in series with the input of G_{m2}, but it is not corrected because the $G_{m2}R$ feedback loop is now open. Thus, ΔV is multiplied by

$G_{m2}R$ and divided by $G_{m1}R$ when referred to the main input. The total input offset voltage is then equal to

$$V_{OS(tot)} = \frac{V_{OS,1}}{G_{m2}R} + \frac{V_{OS,2}}{G_{m1}R} + \frac{G_{m2}}{G_{m1}} \Delta V. \tag{8.9}$$

The three terms in (8.9) impose different constraints on the design of the op amp. To suppress the first two terms, G_{m1}, G_{m2}, and R must be maximized, whereas to reduce the last term, G_{m2}/G_{m1} or ΔV must be minimized. In a typical design, $G_{m2}/G_{m1} \approx 0.1$ and ΔV is reduced using large offset storage capacitors [12].

The topology of Figure 8.10 is easily realized in a folded cascade op amp, with the G_{m1} and G_{m2} output current summation occurring at the folding point. Such an implementation is shown in Figure 8.11, where differential pairs M_1-M_2 and M_3-M_4 comprise the G_{m1} and G_{m2} amplifiers, respectively, and common-gate transistors M_5 and M_6 along with their loads form the R amplifier. The ratio of G_{m1} and G_{m2} is set by the sizing and biasing of M_1-M_4.

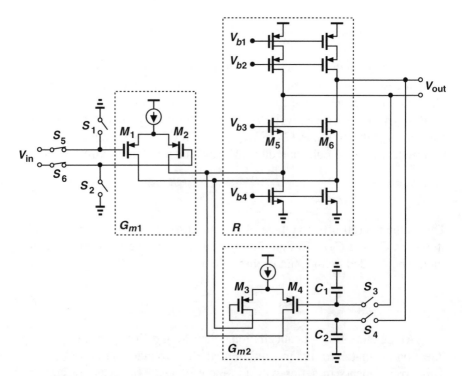

Fig. 8.11   Circuit implementation of topology of Figure 8.10 (common-mode feedback not shown).

Equation (8.9) illustrates the direct dependence of the residual offset upon the loop gains $G_{m1}R$ and $G_{m2}R$, predicting design issues in scaled technologies. For typical short-channel devices, loop gains are smaller, while offsets tend to be larger. Thus, the residual offset of this topology increases for small-geometry devices. Since the supply voltage of such circuits must also scale with technology, the input dynamic range of the op amp is more limited.

8.3 CALIBRATION TECHNIQUES

Integral linearity of data converters usually depends on the matching and linearity of integrated resistors, capacitors, or current sources, and it is typically limited to approximately 10 bits with no calibration. For higher resolutions, means must be sought that can reliably correct nonlinearity errors. This is often accomplished by either improving the effective matching of individual devices or correcting the overall transfer characteristics.

8.3.1 DAC Calibration Techniques

Capacitor DACs.   As mentioned in Chapter 3, capacitor DACs achieve both a wide dynamic range and a relatively fast settling and are often utilized in multistep A/D converters. These DACs have been especially popular in successive approximation architectures, where they are normally configured as binary-weighted arrays.

We first describe a technique for measuring the mismatch between two capacitors. Consider the circuit of Figure 8.12. First, S_1, S_2, and S_5 are on, and hence C_1 is charged to V_{REF} and C_2 is discharged to ground. Next, S_1, S_2, and S_5 turn off and S_3 and S_4 turn on so that nodes X and Y experience voltage swings equal to $-V_{\text{REF}}$ and $+V_{\text{REF}}$, respectively. The change in the voltage at node P, here called the "residual voltage," can be written as

$$\Delta V = -\frac{V_{\text{REF}}}{C_1 + C_2}C_1 + \frac{V_{\text{REF}}}{C_1 + C_2}C_2 \tag{8.10}$$

$$= \frac{C_2 - C_1}{C_1 + C_2}V_{\text{REF}}, \tag{8.11}$$

where the charge injected by S_5 is neglected. Thus, ΔV is a measure of the mismatch between C_1 and C_2.

A calibration technique described by Lee [13, 14] is based on measuring and digitizing the residual voltages obtained for a binary capacitor array and storing and combining the results digitally. Illustrated in Figure 8.13 in simplified form, this approach employs a main DAC, a calibration DAC (CDAC),

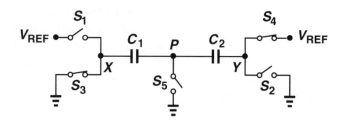

Fig. 8.12   Measurement of mismatch between two capacitors.

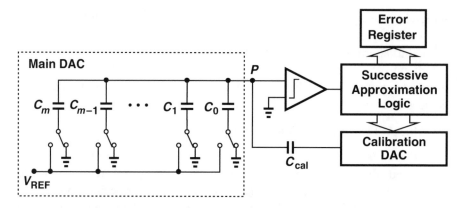

Fig. 8.13   Calibration of a binary-weighted capacitor array.

a precision comparator, and digital logic required for storage and correction. In the main DAC, $C_j = 2^{j-1}C_1$ for $j > 1$, and capacitor C_0, equal to C_1, is used only during calibration.

Before describing the algorithm utilized in this approach, we should mention two points regarding the circuit of Figure 8.13. First, for perfectly matched devices,

$$C_k = \sum_{j=0}^{k-1} C_j, \qquad \text{for } 1 < k \le m, \qquad (8.12)$$

an important result used to measure capacitor mismatches. Second, it can be easily proved that the accuracy of the following algorithm is not influenced by C_{cal}, input capacitance of the comparator, or top plate parasitics of the array.

Calibration begins with C_m and proceeds as follows. Using (8.11) and (8.12), the mismatch between C_m and $C_{m-1} + \cdots + C_0$ is represented as a residual voltage ($V_{\text{res},m}$) at node P. Subsequently, the comparator and CDAC perform a successive approximation routine to digitize $V_{\text{res},m}$. This procedure is repeated for the mismatch between C_j and $C_{j-1} + \cdots + C_0$ for $j = m - 1, \ldots, 1$, thus producing the digital representation of $V_{\text{res},m-1}, \ldots, V_{\text{res},1}$.

For a digital input $D_m D_{m-1} \cdots D_1$, it can be shown [14] that the output error voltage in the presence of capacitor mismatch is

$$V_{\text{error}} = \frac{V_{\text{REF}}}{2^m} \sum_{j=1}^{m} 2^{j-1} \left(\frac{\Delta C}{C}\right)_j D_j \tag{8.13}$$

$$= \sum_{j=1}^{m} V_{\epsilon j} D_j, \tag{8.14}$$

where $(\Delta C/C)_j$ denotes the relative mismatch of C_j and

$$V_{\epsilon j} = \frac{V_{\text{REF}}}{2^m} 2^{j-1} \left(\frac{\Delta C}{C}\right)_j. \tag{8.15}$$

Thus, to find V_{error} for each input code, all $V_{\epsilon j}$ must be calculated. It can also be shown that

$$V_{\epsilon j} = \frac{1}{2}\left(V_{\text{res},j} - \sum_{k=j+1}^{m} V_{\epsilon k}\right). \tag{8.16}$$

Thus, after every $V_{\text{res},j}$ is digitized, a corresponding $V_{\epsilon j}$ is calculated from (8.16) and stored in the data register. The precise error voltage at the output for any digital input can then be determined using (8.14).

The primary feature of this algorithm is that it performs calibration using only adders and simple logic gates, with no need for digital multipliers. In a typical design, the digital circuitry amounts to approximately 400 gates and 120 bits of RAM [13]. The precision achievable through this technique has made possible resolutions as high as 18 bits [15].

Another approach to calibrating capacitor DACs measures the mismatch between each unit capacitor and a "reference" capacitor and corrects the mismatch by adding small, trimmable capacitors to each unit [17]. This approach is better described in the context of ADC calibration and is treated in Section 8.4.

The above calibration techniques fail to fully correct one type of error: capacitor nonlinearity. For example, the mismatch measurement illustrated in Figure 8.12 corrects capacitor nonlinearity only for a capacitor voltage of V_{REF} because the residual voltage also includes the mismatch between the value of C_1 with a voltage of V_{REF} and the value of C_2 with a voltage of zero. However, the nonlinearity corresponding to other voltages across the capacitors remains.

If the capacitor nonlinearity is reproducible and well-characterized, it is possible to employ nonlinear function generators that introduce an opposite nonlinearity in the input/output characteristic, hence canceling the effect of the capacitor voltage coefficient [8]. However, it is desirable to avoid the complexity associated with this technique through the use of fully differential architectures and highly linear capacitors.

Current-Steering DACs.   In order to achieve integral linearities above 10 bits, it is often necessary to calibrate the current sources of a DAC. Factory calibration techniques such as laser wafer trimming and fusing are extensively used in industry, but they increase the cost and their precision degrades with package stress and temperature variation. In contrast, self-calibration techniques are free from these errors but result in more circuit complexity.

The equivalent matching of current sources can be substantially improved using the concept of "dynamic element matching" [16]. Depicted in Figure 8.14, this method performs time averaging on nominally equal current sources I_{R1} and I_{R2} so as to produce precisely matched currents $I_{out,1}$ and $I_{out,2}$. We note that if CK has a duty cycle of exactly 50%, then $I_{out,1} = I_{R1}$ and $I_{out,2} = I_{R2}$ for half of the time and $I_{out,1} = I_{R2}$ and $I_{out,2} = I_{R1}$ for the other half. Thus, both $I_{out,1}$ and $I_{out,2}$ have an average value of $(I_{R1} + I_{R2})/2$ and are therefore exactly equal. In reality, the clock duty cycle may deviate from 50% by some amount, δ_{CK}, thereby yielding a total output mismatch of

$$\frac{\Delta I_{out}}{I_{out}} = \delta_{CK} \frac{I_{R1} - I_{R2}}{I_R}, \qquad (8.17)$$

where I_R is the mean value of I_{R1} and I_{R2} [16]. Since δ_{CK} can be less than 1%, this approach improves the matching by more than two orders of magnitude.

As dynamic element matching is based on time averaging, it requires low-pass filtering of the output currents, e.g., by means of capacitors connected from nodes X and Y in Figure 8.14 to ground. However, unless clock frequencies of several hundred megahertz are used, these capacitors tend to be excessively large for on-chip fabrication, an issue that becomes more severe as the number of averaged outputs increases.

A self-calibration technique for segmented current-steering arrays has been introduced by Groenveld et al. [18]. Illustrated in Figure 8.15, this technique makes each unit current source equal to a reference current source, thus canceling errors due to mismatches. Each unit in the array consists of a fixed current source, M_1, an adjustable current source, M_2, and a hold capacitor, C_H. During calibration, S_1 and S_2 are on and the gate-source voltage of M_2 is adjusted so that $I_1 + I_2 = I_{REF}$. When S_1 and S_2 turn off, this voltage is stored on C_H, retaining the same I_2.

The key point in this circuit is that typical device matching allows M_1 to provide approximately 99% of the total current, thereby relaxing the precision

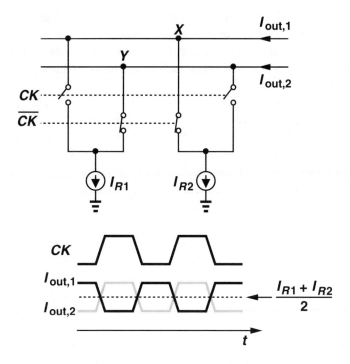

Fig. 8.14   Dynamic element matching.

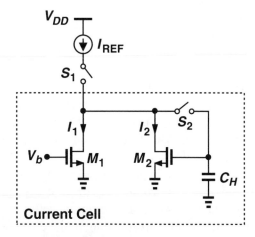

Fig. 8.15   Calibration of a current source.

required of I_2 by two orders of magnitude. For example, in a 16-bit system, calibration must trim I_2 to only 10 or 11 bits of precision, making the circuit quite tolerant of errors due to charge injection and feedthrough. Furthermore, since I_2 is quite small, M_2 can be a long device to attain better matching and lower sensitivity to charge injection at its gate.

In order to make the calibration transparent to the user, a proxy current cell can be added to the DAC array such that it replaces the cell undergoing calibration [18].

The circuit of Figure 8.15 nonetheless has two drawbacks. First, since the adjustable current source has single-ended control (i.e., the gate of M_2), it requires a large value for C_H to suppress the effect of feedthrough due to S_2. If this effect were identical for *all* the cells in the array, then it would simply yield a constant offset in each source and deviate the full-scale current from its ideal value, an error that could be corrected by adjusting I_{REF}. However, if random mismatches and gradients across the chip are significant, feedthrough effects are not identical, thereby causing nonlinearity. To mitigate this problem, the adjustable current source can incorporate differential control.

The second drawback is that if $I_1 > I_{REF}$, calibration fails, implying that I_{REF} must be skewed to guarantee $I_1 \leq I_{REF}$. However, since I_1 and I_{REF} are generated using nonidentical devices (e.g., NMOS and PMOS devices, respectively), it is difficult to ensure that I_1 is less than I_{REF} by only 1%. A more practical approach is to employ a bidirectional current source for I_2 so that it can contribute either negative or positive components to $I_1 + I_2$.

Figure 8.16 shows a calibration circuit that satisfies both of the above requirements. Here, the transconductance amplifier comprising M_2-M_5 and I_{SS} operates as a bidirectional current source to generate I_2. In the calibration mode, S_1-S_3 are on and $I_{D1} + I_{D3} - I_{D5} = I_{REF}$. When S_2 and S_3 turn off, only their charge injection *mismatch* introduces a small error.

Groenveld describes another implementation of the calibration circuit [18].

Another approach described by Miller et al. [19] stores the errors digitally to achieve stable operation over time and temperature without the need for frequent calibration. Depicted in Figure 8.17, this technique employs a main DAC (MDAC), a calibration DAC (CDAC), and a sampling DAC (SDAC) along with control logic and a small memory. In the calibration mode, the output corresponding to each code is examined and the error is detected and stored. This operation occurs in two steps. In the first step, an input code D is applied to the MDAC and the corresponding current cells are switched to node X. In the ideal case, the output corresponding to the next code $D + 1$ would result simply from drawing an additional LSB current from node X. Thus, an *extrapolated* value for $D + 1$ is established at the output by switching an extra LSB current source, I_{LSB}, to X. Subsequently, the SDAC and the comparator operate as a successive approximation ADC and reduce the voltage at node

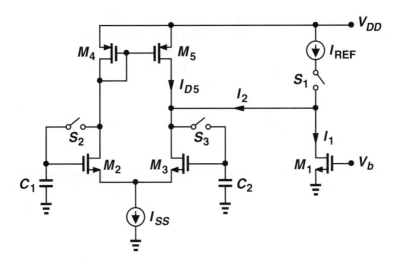

Fig. 8.16   Differential bidirectional calibration of a current source.

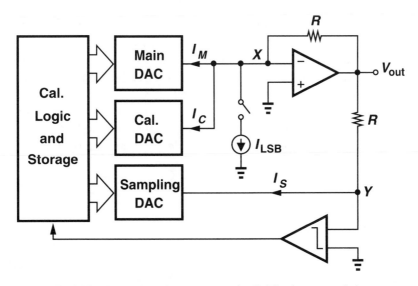

Fig. 8.17   Calibration of a current-steering DAC using extrapolation.

Y to zero. In other words, the SDAC samples the (extrapolated) output of the main DAC.

In the second step, I_{LSB} turns off, and the digital input of the MDAC is incremented by 1 LSB to generate the actual output corresponding to $D + 1$. Next, the CDAC and the comparator successively reduce the voltage of node

Y to zero, making V_{out} equal to the value stored in SDAC. As a result, the difference between the actual output and the extrapolated output is detected by CDAC and stored digitally. This difference is then used to correct the output during actual digital-to-analog conversion.

Since the calibration and sampling DACs only measure or correct *errors*, their resolution and dynamic range can be quite low. For example, in a 16-bit system, these DACs require a resolution of approximately 6 bits [19].

8.3.2 ADC Calibration Techniques

Since high-resolution A/D converters typically employ multistep archi-tectures, they often impose two stringent requirements: small INL in their interstage DACs and precise gain (usually a power of 2) in their interstage subtractors/amplifiers. These constraints in turn demand correction for device mismatches if resolutions above 10 bits are required.

In this section, we describe three calibration techniques suited to mul-tistage ADC architectures. For simplicity, we consider single-ended imple-mentations, but the actual circuits are fully differential.

Capacitor Error Averaging. In the 1-bit-per-stage architecture of Sec-tion 6.6, rather than subdividing the reference, each stage amplifies the residue by a precise factor of 2. The accuracy of this gain depends on capacitor match-ing (and op amp gain). In order to suppress the effect of capacitor mismatch, capacitor error averaging can be used [9, 20].

Figure 8.18 depicts a multiply-by-2 circuit in two modes. In the sampling mode C_1 and C_2 are charged to V_{in}. In the amplification mode, C_1 is placed around the amplifier while node X is grounded. For an ideal op amp,

$$V_{out,1} = V_{in}(1 + \frac{C_2}{C_1}). \qquad (8.18)$$

If $C_1 = C$ and $C_2 = C + \Delta C$, then

$$V_{out,1} = V_{in}(2 + \frac{\Delta C}{C}), \qquad (8.19)$$

thus exhibiting a gain error of $\Delta C/C$, typically on the order of 0.1%.

Now suppose the multiply-by-2 function is repeated but with C_1 and C_2 interchanged. This case is illustrated in Figure 8.19. Then, the output is

$$V_{out,2} = V_{in}(1 + \frac{C_1}{C_2}). \qquad (8.20)$$

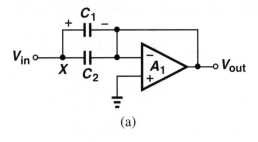

(a)

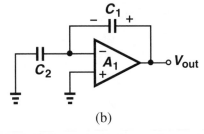

(b)

Fig. 8.18   A multiply-by-2 stage in (a) the sampling mode and (b) the amplification mode.

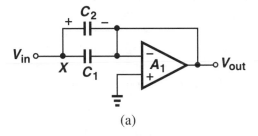

(a)

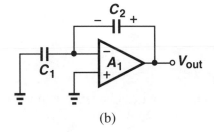

(b)

Fig. 8.19   Multiply-by-2 stage of Figure 8.18 with C_1 and C_2 interchanged.

In the capacitor error-averaging technique, the outputs represented by (8.18) and (8.20) are sampled and averaged to generate

$$V_{out,av} = \frac{1}{2}V_{in}(2 + \frac{C_1}{C_2} + \frac{C_2}{C_1}) \tag{8.21}$$

$$\approx \frac{1}{2}V_{in}(2 + \frac{C_1^2 + C_2^2}{C_1 C_2}). \tag{8.22}$$

We can show that this approach reduces the gain error to a term proportional to the *square* of the relative capacitor mismatch. Substituting $C_1 = C$ and $C_2 = C + \Delta C$ in (8.22), we have

$$V_{out,\ av} = \frac{1}{2}V_{in}[2 + \frac{2C^2 + 2C\,\Delta C + \Delta C^2}{C(C + \Delta C)}] \tag{8.23}$$

$$\approx V_{in}(2 + \frac{\Delta C^2}{2C^2}). \tag{8.24}$$

Thus, the gain error can be expressed as

$$\Delta A_v \approx \frac{\Delta C^2}{2C^2}, \tag{8.25}$$

which is typically only a few parts per million.

Figure 8.20 illustrates the amplifier along with an averaging circuit consisting of A_2, C_3, and C_4. Capacitor C_3 is nominally half of C_4 so as to provide an overall gain of 0.5 for the averaging circuit [9]. During the first amplification mode [Figure 8.20(a)], the output of A_1 is equal to $V_{in}(1 + C_1/C_2)$ and is stored on C_3 and C_4. In the second amplification mode [Figure 8.20(b)], C_4 is placed in a negative feedback loop around A_2, and C_1 and C_2 are interchanged so that the output of A_1 is equal to $V_{in}(1 + C_2/C_1)$. Thus, $V_{out,av}$ is equal to the initial voltage across C_4 plus half of the change in the output of A_1:

$$V_{out,av} = -V_{in}(1 + \frac{C_1}{C_2}) - \frac{1}{2}V_{in}(\frac{C_2}{C_1} - \frac{C_1}{C_2}) \tag{8.26}$$

$$= -V_{in} - \frac{1}{2}(\frac{C_2}{C_1} + \frac{C_1}{C_2})V_{in}. \tag{8.27}$$

The negative sign in the output can be taken into account by the following stage.

Capacitor Trimming.   Figure 8.21(a) shows a 2-bit stage of a pipelined ADC [17] comprising nominally equal capacitors C_1-C_4 and C_F, and op amp

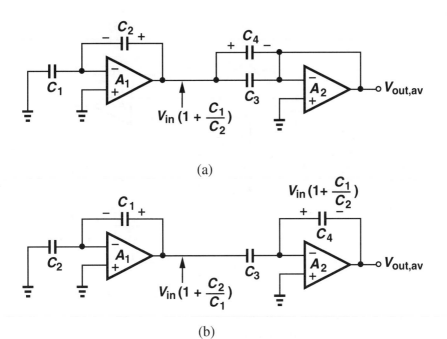

(a)

(b)

Fig. 8.20   Capacitor error-averaging circuit.

A_1. During sampling, S_1-S_4 are switched to V_{in} and S_F is on, providing a virtual ground at the inverting input of A_1. Subsequently, D/A conversion, subtraction, and amplification take place when S_F turns off and S_1-S_4 switch to either V_{REF} or ground according to the thermometer code generated by the previous stage. The output voltage V_{out} is then equal to 4 times the difference between V_{in} and the output of the previous stage.

   In the circuit of Figure 8.21(a), mismatch among C_1-C_4 results in integral nonlinearity, while mismatch between each of C_1-C_4 and C_F causes gain error. In order to reduce these errors, each of C_1-C_4 can be compared with C_F and trimmed so that the effective matching is improved. Illustrated in Figure 8.21(b), the calibration scheme utilizes the (offset-canceled) op amp A_1 as a precision comparator and proceeds as follows. First, the mismatch between C_j ($j = 1, \ldots, 4$) and C_F is represented as a voltage at node X in the same fashion as illustrated in Figure 8.12. Next, the comparator detects the polarity of this voltage and adjusts the value of C_j in the proper direction. This sequence is repeated until the difference between C_j and C_F is reduced to very small values.

   In practice, it is difficult to implement small, trimmable capacitors because, on the one hand, they must be approximately 500 to 1000 times less

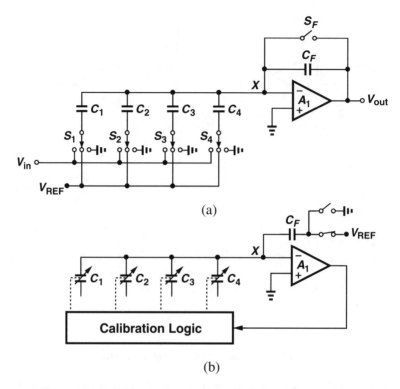

Fig. 8.21 (a) A multibit stage in a pipelined ADC; (b) calibration of the stage using trimmable capacitors.

than C_j and, on the other hand, they must offer sufficient trimming range. For example, for $C_j = 1$ pF, a mismatch of 0.1%, and an overall resolution of 12 bits, the trimmable capacitors will be on the order of a few femtofarads and, more importantly, they must be adjustable in steps of a few tenths of a femtofarad. In current technology, such small capacitors are difficult to fabricate and typically exhibit fringe parasitics that introduce significant errors in their values.

This problem can be overcome through the use of structures consisting of larger capacitors that have a small *equivalent* capacitance. For example, Lin et al. [17] use capacitor ladders for this purpose.

Digital Calibration.   The ADC calibration techniques described above are based on analog processing and correction of nonidealities. Alternatively, calibration can be performed in the digital domain to simplify the measurement and storage of errors.

A digital calibration technique has been proposed for two-step ADCs that employ switched-capacitor interstage D/A conversion and subtraction

[21]. This method computes the additional charge that must be pumped onto the DAC output to correct the reconstructed analog signal and digitally stores the voltage corresponding to that charge.

We consider the 3-bit topology shown in Figure 8.22 to illustrate this approach. This configuration comprises the interstage DAC and subtractor/amplifier of a two-step ADC along with a calibration DAC (CDAC) and a calibration capacitor (C_{cal}). In this circuit, $C_3 = 2C_2 = 4C_1 = 4C_F$. During sampling, S_F is on, establishing a virtual ground at X, and S_1-S_4 are switched to V_{in}. During D/A conversion/subtraction, S_F is off, C_F is placed in a negative feedback loop around A_1, and S_1-S_3 switch to either V_{REF} or ground according to the binary output of the first flash stage. As a result, if all the devices are ideal, the amplifier output is

$$V_{out} = (1 + \frac{C_3 + C_2 + C_1}{C_F})V_{in} - \frac{D_3 C_3 + D_2 C_2 + D_1 C_1}{C_F}V_{REF}, \quad (8.28)$$

where $D = D_3 D_2 D_1$ is the binary output of the previous stage. [Note that the analog equivalent of D is given by $(2^2 D_3 + 2^1 D_2 + 2^0 D_1)V_{REF}/2^3$].

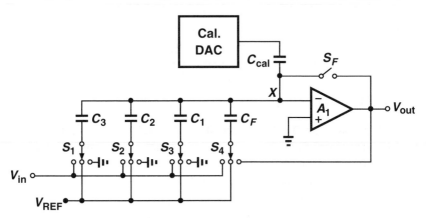

Fig. 8.22   ADC calibration by pumping charge onto interstage DAC output
                 node.

For perfectly matched capacitors, (8.28) reduces to $2^3(V_{in} - 2^{-3}D V_{REF})$, indicating an output equal to 2^3 times the residue. The technique to be described corrects the coefficient of V_{REF} in (8.28) such that it approaches the ideal value of $2^{-3}(2^2 D_3 + 2^1 D_2 + 2^0 D_1)$ and hence improves the linearity of interstage D/A conversion. For simplicity, we assume the open-loop gain of A_1 is sufficiently high.

The calibration is performed for each digital input code and proceeds as follows. First, S_F is on, C_F is charged to V_{REF}, and C_3-C_1 are discharged to ground ($D_3 D_2 D_1 = 000$). Next, C_F is placed around A_1 so as to generate an output of V_{REF}, and the digital input is incremented to $D_3 D_2 D_1 = 001$, i.e., S_3 switches from ground to V_{REF}. If C_1 and C_F match perfectly, then the output goes from V_{REF} to 0. In reality, the mismatch between these capacitors results in a small voltage at the output. To null this voltage, a comparator repeatedly senses the output and drives the CDAC in proper direction so that the charge pumped by C_{cal} onto X balances the extra charge due to mismatch between C_F and C_1, thus reducing V_{out} to zero. The code thus produced by the CDAC is stored in a RAM and applied to CDAC and C_{cal} whenever the digital input from the previous stage switches S_3 to V_{REF}.

In summary, this technique is based on the fact that if the digital input of the DAC increments by 1, the amplifier output should change by exactly V_{REF} and any deviations from this value can be nulled using a calibration DAC and a capacitor.

The calibration technique depicted in Figure 8.22 suffers from accumulation of errors during calibration. This problem can be alleviated using other circuit techniques [21].

In addition to the ADC calibration techniques described in this section, the capacitor DAC calibration of Section 8.3.1 can be employed in successive approximation converters to achieve high linearity [15]. Other digital calibration techniques have been proposed by Karanicolas [22] and Lee [23].

8.4 RANGE OVERLAP AND DIGITAL CORRECTION

A powerful technique to relax the speed-resolution trade-off of multistep A/D converters is to provide overlap between the quantization range of successive stages and digitally correct the binary data produced by each stage. We use a two-step half-flash architecture to illustrate this approach.

Consider the simple two-step 10-bit ADC depicted in Figure 8.23, where each stage resolves 5 bits. Let us first assume an ideal system where comparator offsets are zero, the reference voltages generated by the first-stage ladder are precisely 32 LSB apart, and the quantization range of the second stage is 32 LSB wide. If the input voltage lies between, for example, V_j and V_{j+1}, then the DAC generates a voltage equal to V_j and the subtractor produces a residue of $V_{\text{in}} - V_j$. Since this residue is always less than 32 LSB, it never exceeds the input range of the second stage and hence no gross errors result. This is shown graphically in Figure 8.24(a).

Now consider a nonideal system where, for example, the first-stage comparators exhibit finite offset. In particular, suppose that in Figure 8.23 comparators A_j and A_{j+1} have offsets $\Delta V_j < 0$ and $\Delta V_{j+1} > 0$, respectively.

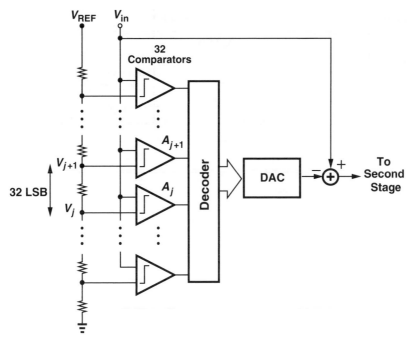

Fig. 8.23   Residue generation in a 10-bit two-step A/D converter.

Consequently, for $V_j + \Delta V_j < V_{in} < V_{j+1} + \Delta V_{j+1}$, A_j generates a ONE output and A_{j+1} a ZERO output; hence the DAC output remains equal to V_j. Thus, as shown in Figure 8.24(b), for $V_j + \Delta V_j < V_{in} < V_j$ the difference between V_{in} and V_{DAC} is *negative*, and for $V_{j+1} < V_{in} < V_{j+1} + \Delta V_{j+1}$, this difference *exceeds* the input range of the second stage. In the first case, the second stage interprets the residue as zero, and in the second as 32 LSB, regardless of its actual value. The resulting input/output characteristic of the overall ADC is shown in Figure 8.24(c), where the dead bands $V_j + \Delta V_j < V_{in} < V_j$ and $V_{j+1} < V_{in} < V_{j+1} + \Delta V_{j+1}$ give rise to maximum differential nonlinearities of ΔV_j and ΔV_{j+1}, respectively.

In addition to offset of the first-stage comparators, several other non-idealities can lead to the overrange problem described above. The reader can prove that the following are among them: nonuniformity in the references generated by the first-stage ladder (if the DAC does not use the same ladder); DAC and subtractor gain error; and long hold-mode settling time of the front-end SHA. This list indicates that, for the ADC of Figure 8.23 to attain 10-bit resolution, the first-stage ladder and comparators, the interstage DAC and subtractor, and the second stage must all achieve a precision of at least 10 bits. Furthermore, the first stage may be strobed only after the SHA output

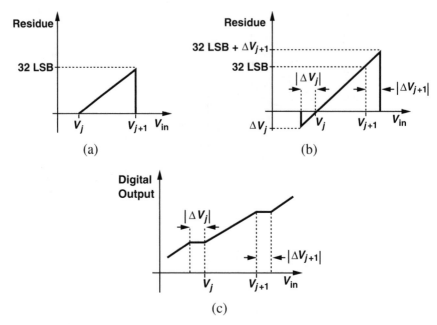

Fig. 8.24   Residue plot for ADC of Figure 8.23. (a) Ideal system; (b) finite offsets in comparators A_j and A_{j+1}; (c) ADC input/output
characteristic in the presence of offsets.

has settled to 10-bit accuracy. In summary, almost all parts of the system
require high precision.

The above constraints all originate from the fact that the second-stage
in Figure 8.23 digitizes residues only between 0 and 32 LSB. Now, as an
example, suppose that the second stage accommodates inputs from −16 LSB
to +48 LSB, i.e., it employs 64 comparators with references ranging from −16
LSB to +48 LSB. Then, these constraints are significantly relaxed because
the total error can be as large as ±16 LSB without introducing overrange, and
only the second-stage comparators require 10-bit resolution. This addition of
redundancy to the second stage is called "overlap," and since the second stage
has one additional bit of resolution, we say the overlap is 1 bit.

For the ADC of Figure 8.23, the overlap between the two stages can be
implemented in one of two forms: (1) if reference voltages from −16 LSB
to +48 LSB are available, the second-stage comparators can simply compare
the residue with these references. This is particularly easy in fully differential
architectures because a reference of +16 LSB can be interpreted as −16 LSB
by exchanging its two terminals; (2) the DAC output can be shifted *down* by
16 LSB so that the residue is always positive (for errors less than 16 LSB).
These cases are depicted in Figure 8.25.

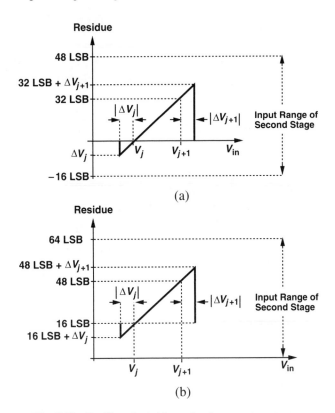

Fig. 8.25   Residue plot with overlap between two stages.

To gain more insight, we discuss an example using the technique shown in Figure 8.25(b). For simplicity, we assume a resistor-ladder DAC, but the technique can be applied to other DAC configurations as well [24]. Figure 8.26 shows a 10-bit ADC using a resistor-ladder DAC. Note that the parallel resistors at the two ends of the DAC ladder shift its tap voltages by 16 LSB, thereby generating the midpoints of the voltage segments produced by the first-stage ladder. If comparators A_j and A_{j+1} interpret V_{in} to lie between V_j and V_{j+1}, then three possibilities exist: V_{in} is slightly less than V_j, but A_j has a negative offset; V_{in} is indeed between V_j and V_{j+1}; or V_{in} is slightly greater than V_{j+1}, but A_{j+1} has a positive offset. We explain how corrections are made in the first and third cases.

If the input is interpreted to be between V_j and V_{j+1}, the AND gate N_j generates an output of ONE, turning on S_j and providing a DAC output equal to $V_j - 16$ LSB. Thus, so long as various errors in the first stage, the DAC, and the subtractor add up to less than 16 LSB, the residue remains positive, and hence no overrange occurs.

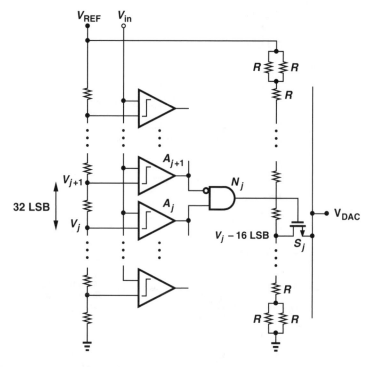

Fig. 8.26   10-bit ADC using a resistor-ladder DAC.

In order for the third case explained above not to introduce overrange errors, the second stage is designed for 6-bit resolution; i.e., it quantizes residues ranging from 1 LSB to 64 LSB. As a consequence, for the typical situation depicted in Figure 8.26, the second stage digitizes the difference between V_{in} and $V_j - (16\,\text{LSB})$ and has no overrange errors if $V_{in} < V_{j+1} + 16$ LSB.

In order to produce the final binary output, the data generated by the two stages must be digitally corrected. If the digital output of the first stage is represented by a binary number $X_1 = D_{10} \ldots D_6$, then its value normalized to the LSB of the 10-bit system is $X_1 = D_{10}2^9 + D_9 2^8 + \cdots + D_6 2^5$. Similarly, the digital output of the second stage, $X_2 = D_5 \ldots D_0$, can be expressed as $X_2 = D_5 2^5 + D_4 2^4 + \cdots + D_0 2^0$. Because the DAC output is shifted down by 16 LSB, the final output is

$$
\begin{aligned}
Y &= (D_{10}2^9 + D_9 2^8 + \cdots + D_6 2^5) - 2^4 \\
&\quad + (D_5 2^5 + D_4 2^4 + \cdots + D_0 2^0).
\end{aligned}
\tag{8.29}
$$

This equation indicates that the digital correction procedure consists of three steps: (1) shift $D_{10} \ldots D_6$ to the left by 5 bits, (2) subtract the binary number

10000 from the result of step 1, and (3) add $D_5 \ldots D_0$ to the difference obtained in step 2.

We should mention two important points. First, the overlap and digital correction techniques described above cannot correct for subtractor gain error. Second, the total amount of overlap between consecutive stages can be more or less than 1 bit, depending on the overall anticipated error.

In the ADC of Figure 8.26, the overlap is implemented by expanding the quantization range of the second stage, resulting in a 5-bit first stage and a 6-bit second stage. Alternatively, the first stage can be designed to have a higher resolution, allowing fewer comparators in the second stage. For example, the ADC can resolve 6 bits in the first stage and 5 bits in the second. The exact allocation of resolution to the stages depends on various issues pertinent to each particular design. For example, in CMOS technology, comparators typically suffer from severe trade-offs among their speed, resolution, power dissipation, and input capacitance. Thus, it is desirable to use fewer high-resolution comparators in the second stage and hence more low-resolution comparators in the first stage. This also relaxes the gain accuracy required of the subtractor. On the other hand, as the resolution of the first stage and hence that of the interstage DAC increases, the DAC may exhibit a longer settling time.

REFERENCES

[1] G. Erdi, "A Precision Trim Technique for Monolithic Analog Circuits," *IEEE J. Solid-State Circuits*, vol. SC-10, pp. 412-416, Dec. 1975.

[2] D. J. Allstot, "A Precision Variable Supply CMOS Comparator," *IEEE J. Solid-State Circuits*, vol. SC-17, pp. 1080-1087, Dec. 1982.

[3] B. Razavi and B. A. Wooley, "Design Techniques for High-Speed High-Resolution Comparators," *IEEE J. Solid-State Circuits*, vol. SC-27, pp. 1916-1926, Dec. 1992.

[4] J. Doernberg, P. R. Gray, and D. A. Hodges, "A 10-Bit 5-Msample/s CMOS Two-Step Flash ADC," *IEEE J. Solid-State Circuits*, vol. SC-24, pp. 241-249, April 1989.

[5] Y. S. Yee, L. M. Terman, and L. G. Heller, "A 1 mV MOS Comparator," *IEEE J. Solid-State Circuits*, vol. SC-13, pp. 294-297, June 1978.

[6] A. G. Dingwall and V. Zazzu, "An 8-MHz CMOS Subranging 8-Bit A/D Converter," *IEEE J. Solid-State Circuits*, vol. SC-20, pp. 1138-1143, Dec. 1985.

[7] K. Kusumoto et al., "A 10 b 20 MHz 30 mW Pipelined Interpolating CMOS ADC," *IEEE J. Solid-State Circuits*, vol. SC-28, pp. 1200-1206, Dec. 1993.

[8] R. K. Hester et al., "Fully Differential ADC with Rail-to-Rail Common Mode Range and Nonlinear Capacitor Compensation," *IEEE J. Solid-State Circuits*, vol. SC-25, pp. 173-183, Feb. 1990.

[9] B. S. Song, M. F. Tompsett, and K. R. Lakshmikumar, "A 12-Bit 1-Msample/s Capacitor-Averaging Pipelined A/D Converter," *IEEE J. Solid-State Circuits*, vol. SC-23, pp. 1324-1333, Dec. 1988.

[10] R. Poujois and J. Borel, "A Low Drift Fully Integrated MOSFET Operational Amplifier," *IEEE J. Solid-State Circuits*, vol. SC-13, pp. 499-503, Aug. 1978.

[11] Y. Tsividis and P. Antognetti, *Design of MOS VLSI Circuits for Telecommunications*, Prentice-Hall, New Jersey, 1985.

[12] E. Sackinger and W. Guggenbuhl, "A Versatile Building Block: The CMOS Differential Difference Amplifier," *IEEE J. Solid-State Circuits*, vol. SC-22, pp. 287-294, April 1987.

[13] H. S. Lee, D. A. Hodges, and P. R. Gray, "A Self-Calibrating 15 Bit CMOS A/D Converter," *IEEE J. Solid-State Circuits*, vol. SC-19, pp. 813-819, Dec. 1984.

[14] H. S. Lee and D. A. Hodges, "Self-Calibration Technique for A/D Converters," *IEEE Trans. Circuits Syst.*, vol. CAS-30, pp. 188-190, March 1983.

[15] G. A. Miller, "An 18b 10 μs Self-Calibrating ADC," *ISSCC Dig. Tech. Papers*, pp. 168-169, Feb. 1990.

[16] R. J. van de Plassche, "Dynamic Element Matching for High-Accuracy Monolithic D/A Converters," *IEEE J. Solid-State Circuits*, vol. SC-11, pp. 795-800, Dec. 1976.

[17] Y. M. Lin, B. S. Kim, and P. R. Gray, "A 13-b 2.5-MHz Self-Calibrated Pipelined A/D Converter in 3-μm CMOS," *IEEE J. Solid-State Circuits*, vol. SC-26, pp. 628-636, April 1991.

[18] D. W. J. Groenveld et al., "A Self-Calibration Technique for Monolithic High-Resolution D/A Converters," *IEEE J. Solid-State Circuits*, vol. SC-24, pp. 1517-1522, Dec. 1989.

[19] G. A. Miller et al., "A True 16b Self-Calibrating BiCMOS DAC," *ISSCC Dig. Tech. Papers*, pp. 58-59, Feb. 1993.

[20] P. W. Lee et al., "A Ratio-Independent Algorithmic Analog-to-Digital Converter Technique," *IEEE J. Solid-State Circuits*, vol. SC-19, pp. 828-836, Dec. 1984.

[21] S. H. Lee and B. S. Song, "Digital-Domain Calibration of Multistep Analog-to-Digital Converters," *IEEE J. Solid-State Circuits*, vol. SC-27, pp. 1679-1688, Dec. 1992.

[22] A. N. Karanicolas, H. S. Lee, and K. L. Bacrania, "A 15b 1MS/s Digitally Self-Calibrated Pipeline ADC," *IEEE J. Solid-State Circuits*, vol. SC-28, pp. 1207-1215, Dec. 1993.

[23] H. S. Lee, "A 12 Bit 600 kS/s Digitally Self-Calibrated Pipeline Algorithmic ADC," *Proc. VLSI Circuits Symp.*, pp. 121-122, May 1993.

[24] B. Razavi and B. A. Wooley, "A 12-b 5-Msample/s Two-Step CMOS A/D Converter," *IEEE J. Solid-State Circuits*, vol. SC-27, pp. 1667-1678, Dec. 1992.

9

Testing and Characterization

The large number of parameters required to fully specify the performance of data conversion interfaces and the presence of both analog and digital signals in these circuits make testing and characterization a lengthy, elaborate task. Furthermore, as these circuits are designed to provide higher levels of performance, their speed and resolution exceed those of typical characterization equipment, often demanding custom-designed measurement systems.

In this chapter, we describe a number of techniques for characterizing data acquisition systems, with emphasis on issues related to high-speed testing. To measure the analog output of sampling circuits and D/A converters, we introduce limiting amplifiers, digitizers, and sinusoidal testing. For A/D converters, we describe downsampling, code density tests, fast Fourier transform tests, and sine fitting. In practice, a combination of these techniques must be used to obtain an accurate and complete set of parameters.

9.1 GENERAL CONSIDERATIONS

The three classes of data acquisition circuits, namely, sampling circuits, A/D converters, and D/A converters, generally require a test setup as shown in Figure 9.1, where one or both of the input and output signals are analog depending on the type of circuit.

In this setup, while the clock usually runs at its maximum rate, the input frequency and amplitude can determine two types of testing: static and

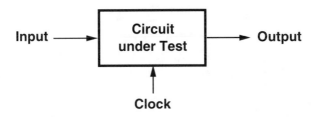

Fig. 9.1   General test setup for data acquisition circuits.

dynamic. In static testing, the input varies slowly and static parameters such as nonlinearity, offset, and gain error are measured. In dynamic testing, the input varies substantially from one clock cycle to the next to reveal the response of the circuit to rapidly changing signals, and dynamic parameters such as signal-to-noise ratio and harmonic distortion are measured. The ultimate dynamic test takes place if the input has a full-scale swing and a bandwidth equal to half the clock rate. These issues are discussed further in the following sections.

While in practice the input can assume any waveform, characterization of the circuit under test must be performed with precise, well-defined inputs that provide useful information for a wide variety of applications. In addition, the input waveform must be easy to generate and also yield outputs that are easy to measure.

For static testing, the input waveform in Figure 9.1 can be simply a ramp that varies from zero to full scale with high linearity [Figure 9.2(a)], ideally producing exactly the same waveform at the output. For dynamic testing, on the other hand, a realistic high-speed *analog* ramp appears as shown in Figure 9.2(b), where the waveform suffers from rounding and finite falltime due to the limited bandwidth of the system as well as ringing due to various sources of resonance in the setup. Since in all three classes of data acquisition circuits the input and/or the output will be an analog ramp, it becomes increasingly difficult to isolate and quantify the nonidealities of the circuit and those of the test setup simply by examining the distorted ramp output.

Sinusoidal waveforms are more suitable for dynamic testing. These functions have several properties that make them attractive for reliable, accurate testing. First, they have a precise mathematical definition in both time and frequency domains and approach their ideal form by proper filtering (for analog) or increasing the word length (for digital). Second, their nonidealities, i.e., distortion and noise, can be easily measured by means of spectral analysis. Third, they provide a single frequency component that can be varied to reach half of the clock rate, thereby giving accurate information about the circuit's frequency response. Fourth, in linear systems, the output response to a complex input waveform can be expressed as the sum of responses to the

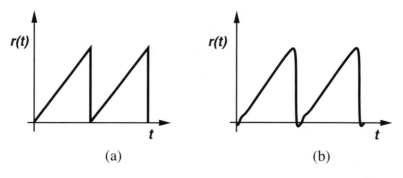

Fig. 9.2   (a) Ideal and (b) actual ramp waveforms.

sinuoidal components comprising that input. Consequently, as we will see throughout this chapter, sinusoidal testing is widely used for all data acquisition interfaces.

9.2 SAMPLING CIRCUITS

The principal difficulty in characterizing sample-and-hold amplifiers arises from the fact that both the input and the output are analog signals that must be measured with sufficient precision. We describe measurement techniques for a number of SHA parameters.

In order to measure the worst-case acquisition and hold-mode settling times (t_{aq}, t_{hs}, respectively), the analog input can be switched from zero to full-scale voltage in consecutive clock cycles such that the circuit experiences the largest possible transients. Figure 9.3 illustrates the waveforms in such a measurement. The analog input is a fast-settling rectangular wave that has a frequency half that of the clock. At $t = t_1$, the circuit enters the acquisition mode and the output begins to swing from zero to full-scale. The circuit enters the hold mode at $t = t_2$, and the analog input goes from full-scale to zero before t_3. At $t = t_3$, the circuit begins to acquire the next sample.

In practice, several issues exist in this measurement. First, any ringing on the analog input may appear at the output, thus increasing the measured acquisition time. For this reason, the analog input must be examined and timed carefully to ensure that it settles before the circuit enters the acquisition mode. Second, delays associated with clock and output signal paths in a typical test setup can introduce errors in t_{acq} and t_{hs} because these two parameters are measured with respect to the clock edge. Thus, the delays must be accurately measured and taken into account.

The third and perhaps the most important problem in the measurement technique of Figure 9.3 relates to the wide dynamic range and fast recovery

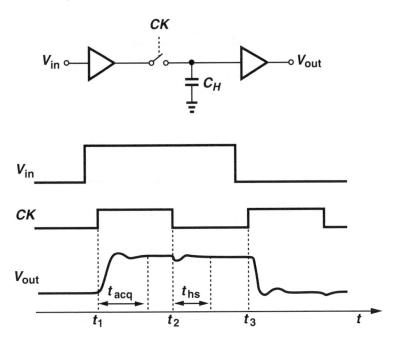

Fig. 9.3   Waveforms for testing sampling circuits.

required of the *instrument* that senses and displays the SHA output. For example, in a typical oscilloscope the input vertical amplifier has a dynamic range hardly exceeding 10 bits; e.g., it is overdriven heavily if the sensitivity is set to 1 mV per division (to display settling behavior) while the overall swing is greater than 1 V. When the input amplifier is overdriven, it may take a long time to recover, thus corrupting the acquisition time measurement. Furthermore, since vertical amplifiers are typically designed for accuracies of no more than 10 bits, little attention is paid to their settling to higher precisions, precluding conclusive measurements for resolutions above 10 bits.

   In order to alleviate the overdrive problem, a limiting amplifier can be interposed between the SHA and the oscilloscope such that it masks most of the SHA output swing and amplifies only the settling portion of the waveform. Shown in Figure 9.4(a), this approach in effect "magnifies" the settling behavior after the SHA output passes a certain level, V_{REF}. The limiting amplifier must of course exhibit sufficiently fast overdrive recovery, but it need not have a wide dynamic range, large output swings, or high linearity. Thus, it can be designed as an open-loop circuit with a low gain and a small output swing so as to provide fast output recovery. A simple example is depicted in Figure 9.4(b), where the differential pair provides the limiting for large differences at its input and emitter degeneration is used to attain sufficient linearity. In

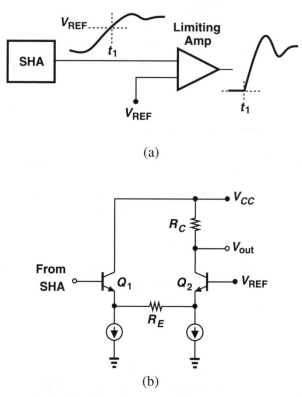

Fig. 9.4   (a) Use of limiting amplifier at the output of a SHA; (b) simple
limiting amplifier.

practice, the gain and offset of the limiting amplifier must be calibrated or
precisely measured and taken into account.

The concept of the limiting amplifier can be extended to a digitizing
technique whereby a comparator is used to sample points on the waveform [1].
If the delay between the SHA clock and the comparator clock is incremented
in small steps and the reference input of the comparator is adjusted properly,
the sampling instant "walks" along the waveform, showing the exact time at
which the signal passes a certain level.

Illustrated in Figure 9.5, the digitizing technique employs a comparator
and an integrator in a feedback loop and a programmable delay generator
that strobes the comparator with a certain delay after the SHA clock. The
comparator output is integrated and fed back to its other input. We explain
the circuit's operation with the aid of the waveforms shown in Figure 9.6. If
at the time of the comparator strobe (when CK_2 goes high), V_{SH} is at V_1, then
V_{com} remains high so long as $V_{int} < V_1$. As V_{com} is integrated, V_{int} approaches

V_1. After V_{int} crosses V_1, the loop enters a pseudostable condition where V_{com} toggles between low and high and V_{int} oscillates around V_1. When this condition occurs, the average integrator output is equal to V_1, i.e., the value of V_{SH} at the comparator strobe instant. This average can be measured using a high-precision voltmeter with little concern for speed. Subsequently, the delay between CK_1 and CK_2 is incremented so that the comparator samples another point from V_{SH} and the above procedure is repeated. If the delay of CK_2 is incremented in sufficiently small steps and the above sequence is repeated for each step, the SHA output waveform can be reconstructed accurately.

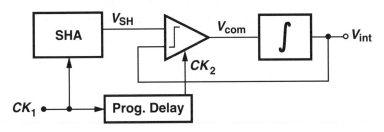

Fig. 9.5   Digitizing technique to measure settling time of SHA output.

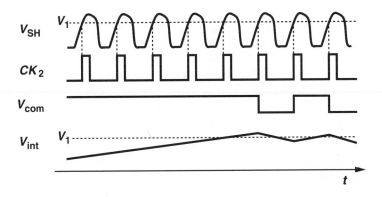

Fig. 9.6   Waveforms used in the setup of Figure 9.5.

Since comparators typically achieve high speeds, they can provide a means for fast sampling in the above approach. Nevertheless, care must be taken so that the comparator kickback noise does not disturb the SHA output at the sampling instant. Also, the comparator offset, which appears in the output as well, must be nulled or precisely measured.

In addition to eliminating the need for wide dynamic range instrumentation, the digitizing technique of Figure 9.5 can also serve as an on-chip tester for circuits that need not drive low-impedance loads. For example, the front-end SHA of an A/D converter is quite difficult to test at full speed individually unless it is (unnecessarily) designed to drive 50-Ω loads. On the other hand, if some of the components of the digitizer such as the comparator and the integrator are built along with the ADC, they can accurately monitor the SHA output with negligible loading.

Another important aspect of SHAs is their linearity and hence output harmonic distortion. The output waveform of a typical sampling circuit consists of "acquisition edges" followed by relatively flat held levels and exhibits two types of nonlinearity. The first type arises from nonlinear phenomena at the beginning of the acquisition mode: for example, the sampling switch on-resistance may vary substantially, thereby introducing an input-dependent delay, or the edges may be slew rate–limited. The acquisition edges therefore give rise to significant nonlinearity in the output waveform. The second type of nonlinearity appears in the settled levels in the acquisition and hold modes and, as discussed in Chapter 2, originates from circuit nonlinearities, input-dependent charge injection, etc.

The distinction between these two types of nonlinearity is necessary because, depending on the application, one or both may be important. If a SHA is used at the front end of an ADC, then its output is sensed only during the hold mode and hence the nonlinearity due to acquisition edges can be neglected. On the other hand, if a SHA is employed as a deglitcher at the output of a DAC, then its entire output waveform is sensed by the following circuit (usually a low-pass filter) and hence both types of nonlinearity become critical. In the first case, only the settled levels in the hold mode should be utilized to calculate the nonlinearity and harmonic distortion [2], whereas in the second case the entire waveform must be considered.

As explained in Section 9.1, the linearity of a SHA can be measured by applying a ramp at the input, but the accuracy of the ramp degrades substantially at high frequencies. As a consequence, sinusoidal inputs are instead used and the output is examined for harmonic distortion content. This can be done by either examining the output on a spectrum analyzer or resampling and digitizing the output using an A/D converter and performing a discrete spectral analysis.

While acquisition time is often quoted as a measure of the speed of SHAs, some sampling circuits do not provide an observable acquisition behavior at their output. For example, in the switched-capacitor circuit of Figure 3.11, the output voltage is fixed at zero during sampling, thus precluding direct measurement of the acquisition time. In such circuits, the speed-precision trade-off can be characterized only by applying sine waves at the input and measuring the output signal-to-(noise + distortion) as a function of the sampling rate and the analog input frequency.

9.3 D/A CONVERTERS

The measurement of static parameters of DACs, such as DNL, INL, gain error, and offset, is relatively straightforward and can be easily automated. On the other hand, characterizing the dynamic behavior of DACs entails a number of issues.

As explained in Chapter 4, to obtain the worst-case settling time of a DAC, the digital input must toggle between zero and full-scale in consecutive cycles so that the analog output experiences the largest change in the shortest time (Figure 4.4). This indicates that measurement of DAC settling times is similar to that of SHA acquisition times; they both require instrumentation having a wide dynamic range. Thus, the digitizing technique described in the previous section can be utilized for DACs to eliminate the need for such instruments.

In another dynamic test, the input is produced by a digital sine wave generator and the analog output is examined for noise and harmonic distortion by means of a spectrum analyzer. Revealing the combined effects of DNL, INL, settling time, and glitch impulse (albeit for only a sinusoidal input), this measurement provides a relatively complete view of the DAC's performance as the input frequency varies.

Digital sine waves can be generated by either programmable pattern generators or integrated circuits specifically designed for "direct digital synthesis" [3].

9.4 A/D CONVERTERS

As with other data acquisition circuits, A/D converters have a large number of parameters that need to be measured. The digital output can be examined as a continuous-time, continuous-amplitude signal if converted to analog form or simply as a digital signal if analyzed using discrete signal techniques.

9.4.1 Static Testing

An automated test often used to determine the static parameters of ADCs is the servo-loop technique, illustrated in Figure 9.7 [1]. The test incorporates a digital comparator, an integrator, and the ADC in a negative feedback loop so as to determine the analog signal level required at the ADC input for every digital code transition.

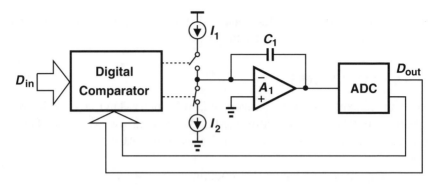

Fig. 9.7   Servo loop used for static characterization of ADCs.

A typical test proceeds as follows. A digital code is applied to one input of the comparator, which causes C_1 to charge (by means of I_1) or discharge (by means of I_2) according to the polarity of the difference between D_{in} and D_{out}. The ADC digitizes the resulting integrator output, producing a D_{out} closer to D_{in}. The loop settles when $D_{out} = D_{in}$. The integrator output voltage then represents the analog level at the ADC input that generates D_{out}. This voltage can be measured using a high-precision voltmeter.

If D_{in} covers the range 0 to $2^m - 1$, where m is the ADC's resolution, an input/output characteristic such as that of Figure 9.8 can be constructed. From this characteristic, various static errors such as offset, DNL, INL, and gain error are computed.

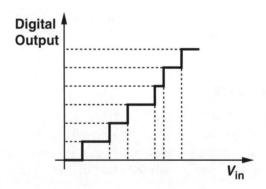

Fig. 9.8   Input/output characteristic produced by a servo-loop measurement.

An important feature of this technique is that it does not require precision ramp generators. For example, the devices comprising the integrator may have substantial nonlinearity without any adverse effect on the measurement accuracy. This is because the digital comparator introduces a virtually infinite gain in the feedback loop, suppressing the integrator nonlinearity.

9.4.2 Dynamic Testing

An important issue in both characterizing and using ADCs is their response to high-frequency *analog inputs*, especially when their amplitude reaches full-scale. For example, an 8-bit, 100-MHz ADC may exhibit 8-bit resolution when digitizing analog inputs of a few megaherz at a conversion rate of 100 MHz. However, the performance of the same converter may degrade drastically as the analog input frequency approaches 50 MHz and the amplitude reaches full-scale.

As explained in Section 9.1, while static ADC tests can utilize ramp waveforms to determine the input/output characteristic, dynamic tests are often performed using sinusoidal inputs.

Direct ADC-DAC Test.   A simple, efficient test procedure using sinusoidal inputs is depicted in Figure 9.9. Here, the ADC under test is followed by a DAC that has sufficiently higher linearity and lower noise than the ADC itself (typically by 2 bits). The DAC output is examined on a spectrum analyzer. If the noise and distortion contributed by the DAC are negligible, the SNR and SNDR measured by the spectrum analyzer correspond to those of the ADC. Since it requires no digital processing of the data, this method allows a quick and easy dynamic testing and provides real-time feedback regarding the circuit's behavior as test conditions are varied.

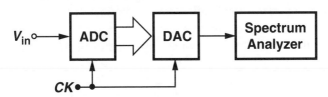

Fig. 9.9   Characterization of an ADC using a high-precision DAC.

An important consideration in using sinusoidal inputs is their frequency relation with the clock, i.e., the ratio of the sine wave period and the clock period. If the ratio of the two periods, T_{in}/T_{CK}, is a rational number, m/n, where m and n are integers with no common divisor, then nT_{in} $(= mT_{CK})$ is an integer multiple of the period of both waveforms, i.e., the analog and clock inputs "beat" every nT_{in} seconds. If, as depicted in Figure 9.10, nT_{in} is small, then only a few points on the sine wave are sampled and some input levels of the ADC remain unexamined. Thus, in order for every code transition to be tested, nT_{in} must be sufficiently large. In practice, this is accomplished by choosing the sine wave frequency as an integer "submultiple" of the clock frequency plus a small offset (e.g., $f_{in} = 0.5f_{CK} + \Delta f$). Note that the relationship $nT_{in} = mT_{CK}$ is guaranteed only if the input sine wave and the clock are locked to the same time base. If, on the other hand, these waveforms are provided by independent generators, their frequencies tend to drift with time, making m/n time-dependent.

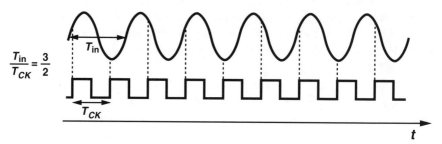

Fig. 9.10   Sinusoidal and clock waveforms with high beat frequency.

If the clock frequencies of the ADC and the DAC in Figure 9.9 are equal, the DAC must achieve a dynamic performance better than that of the ADC so as not to influence the accuracy of measurements. Thus, for testing a state-of-the-art ADC, a high-performance DAC is required. However, the sampling frequencies of the ADC and the DAC need not be equal; if the DAC samples the ADC output only once every M clock cycles, then all the input levels of the ADC are still tested but over a longer period of time. This concept, illustrated in Figure 9.11 and called downsampling, greatly eases the settling speed required of the DAC, thus allowing a high precision in that circuit [4]. In practice, the ADC clock is divided by a factor of M and applied to the DAC.

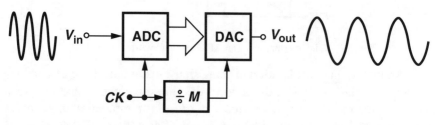

Fig. 9.11   Downsampling.

Beat Frequency and Envelope Tests.   Another test that can be performed using the setup of Figure 9.9 is the beat frequency test wherein the input sine frequency is equal to the sampling frequency f_S plus a small increment Δf [Figure 9.12(a)] [5]. The sampling instants then "walk" along the sine wave at a rate equal to Δf and hence can be reconstructed by a slow, high-precision D/A converter. The reconstructed sine wave provides a quick, qualitative demonstration of the ADC's dynamic performance. If missing codes and sparkles are present, they manifest themselves as shown in Figure 9.12(b).

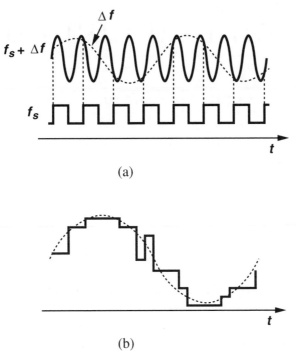

(a)

(b)

Fig. 9.12   (a) Waveforms used in beat frequency test; (b) effect of missing codes and sparkles in the beat output.

In a beat frequency test, it is desirable to set the offset frequency Δf such that successive samples differ by no more than 1 LSB. For a full-scale sinusoidal input, $V_{in} = A \sin \omega t$ (1 LSB $= 2A/2^m$, where m is the ADC's resolution), the offset frequency is equal to $\omega/2^{m-1}$ [6].

The above beat frequency test yields somewhat unrealistic results because it tests ADCs with equal input and sampling frequencies, a condition seldom found in practice. A variant of this test, called the "envelope test" [5], uses an input sine wave with a frequency equal to *half* the sampling rate plus a small increment Δf, thus emulating Nyquist-rate operation (Figure 9.13). If every other sample is considered, then an envelope having a frequency equal to Δf results. Since in this test successive samples differ by as much as the ADC full-scale range, the true dynamic performance of the circuit is revealed.

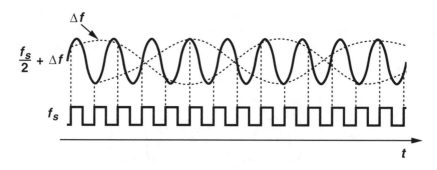

Fig. 9.13   Waveforms used in envelope test.

While the above tests demonstrate some aspects of the performance of ADCs, thorough dynamic characterization often requires that the digital output be collected by means of a memory or a computer and processed mathematically, thus avoiding the need for precision DACs. We describe three tests based on this approach.

Code Density Test.   A common characterization is the histogram or code density test [5, 7], wherein a large number of output samples are collected and their frequency of occurrence is plotted as a histogram versus possible output codes. Such histograms are commonly constructed with "bins," each representing an output code and hence all the *input* values that produce that code. During the test, every time V_{in} is between transition points V_j and V_{j+1}, the bin height of the output corresponding to V_{j+1} is incremented by 1. For example, if an ideal, full-scale ramp is applied to an ideal ADC, the histogram consists of equal-sized bins for all output codes [Figure 9.14(a)] because the input distribution is uniform and the ADC exhibits equal probability of generating any of the codes. On the other hand, if the ADC is not ideal, the bins may not have equal height because some codes occur more often than others [Figure 9.14(b)]. To illustrate this point further, let us consider the ideal and nonideal characteristics in Figures 9.15(a) and (b), respectively, where each input/output plot is accompanied by its histogram (rotated by 90°). A comparison of these two cases shows that the output code corresponding to V_{j+1} occurs more frequently than that corresponding to V_j when a finite differential nonlinearity exists. This gives the important conclusion that the difference between the height of adjacent bins is proportional to the DNL.

In dynamic testing, the code density test employs sinusoidal inputs, and the histogram assumes a different form. If the clock and sine frequencies are chosen such that the sampling instants occur with equal probability on

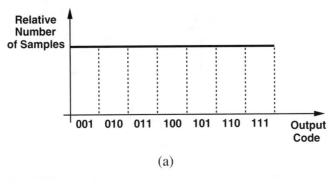

(a)

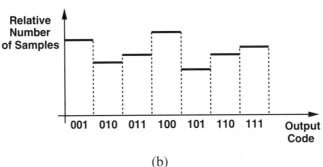

(b)

Fig. 9.14   Output data histogram for an ideal ramp applied to (a) an ideal
ADC and (b) a nonideal ADC.

the time axis, then, in $V_{in} = A \sin \omega t = g(t)$, t can be viewed as a random
variable with a uniform distribution between 0 and T; i.e.,

$$f_t(t) = \frac{1}{T} \quad 0 < t < T \tag{9.1}$$

$$= 0 \quad \text{otherwise}$$

where $f_t(t)$ is the probability density function of t. Since V_{in} is a function of
the random variable t, its PDF can be found using the identity

$$f_{Vin}(V_{in}) = \left| \frac{d}{dV_{in}} g^{-1}(V_{in}) \right| f_t(V_{in}), \tag{9.2}$$

which yields

$$f_{Vin}(V_{in}) = \frac{1}{\pi\sqrt{A^2 - V_{in}^2}}. \tag{9.3}$$

Figure 9.16 plots this distribution. Intuitively, we can say that the points
near zero crossings of the sine wave occur less frequently than those near

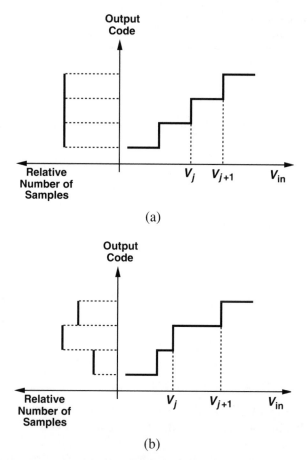

Fig. 9.15   Input/output characteristic and histogram of (a) an ideal ADC and
(b) an ADC with DNL.

the peaks because the signal varies much more rapidly at zero crossings and
hence is less likely to be sampled at those points.

The histogram test easily reveals differential nonlinearity, offset, and
gain error of ADCs. DNL appears as deviation of bin heights from their ideal
value, offset as a horizontal shift, and gain error as horizontal compression
or expansion of the plot. Note that accurate measurement of gain error is
possible only if the input signal amplitude is precisely known and does not
exceed the ADC input range. Figure 9.17(a) illustrates the effect of offset and
differential nonlinearity, exhibiting a missing code when DNL ≤ -1 LSB.

In order to obtain quantitative information, the code density plot can be
normalized with respect to its ideal form given by (9.3). Depicted in Figure
9.17(b) is an example of the resulting plot.

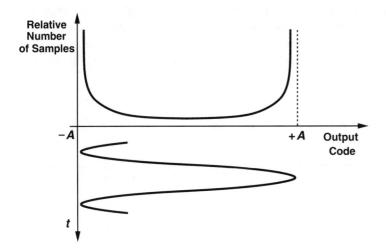

Fig. 9.16 Histogram plot for an ADC with sinusoidal input.

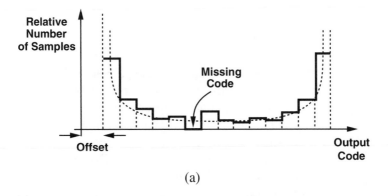

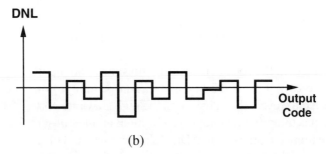

Fig. 9.17 (a) Output histogram of an ADC having large errors; (b) DNL plot calculated from the histogram.

In principle, integral nonlinearity can also be derived from the normalized code density plot. For this purpose, we prove that INL can be obtained from the cumulative sum ("integral") of DNL. Defining $DNL_k = V_{k+1} - V_k - (1 \text{ LSB})$, where V_{k+1} and V_k are consecutive transition points, we can express any transition point V_j as:

$$V_j = \sum_{k=0}^{j}(V_{k+1} - V_k) \tag{9.4}$$

$$= (j \text{ LSB}) + \sum_{k=0}^{j} DNL_k, \tag{9.5}$$

where $V_0 = 0$. Assuming zero gain error and offset for simplicity, we note that the ideal level corresponding to V_j is equal to j LSB, and the INL profile is given by the difference between the actual and ideal characteristics:

$$INL_j = \sum_{k=0}^{j} DNL_k. \tag{9.6}$$

In practice, other errors tend to accumulate during histogram data collection, thereby corrupting the INL calculation described above. These errors result from drift of supply voltage, drift of the input sinusoid amplitude and offset, and drift of the reference voltage and the gain and offset of the ADC itself [7]. The impact of these drifts can be substantial if the data is collected over a long period of time, for example, if the data buffer used at the output is not sufficiently deep and must transmit the data to a computer via a slow link. Consequently, for linearities above 10 bits, code density results may not yield an accurate INL. In that case, the principal effect of INL, namely, harmonic distortion, is measured using other techniques with greater accuracy.

A drawback of the code density technique is that random noise in the ADC under test tends to equalize the bin size of adjacent codes, possibly concealing large DNL errors [8]. This is particularly important in the case of missing codes because random noise simply increases the height of their corresponding bins.

Since in the code density test with sinusoidal inputs most samples occur near the two ends of the histogram, a large number of data points must be collected to reveal the behavior in the high slew rate portion of the waveform. Thus, it is important to know the minimum number of samples required for accurate representation of an ADC's characteristics. This number is determined by statistical significance and "level of confidence," explained in [5, 7].

While histogram testing provides information on each code transition, other methods can give a global assessment of ADCs, primarily in terms of noise and harmonic distortion. Of these, the fast Fourier transform (FFT) and sine fitting techniques are described here.

FFT Test.   The FFT test converts the ADC output from time domain to frequency domain, in a sense similar to the ADC-DAC spectrum analyzer method of Figure 9.9. Thus, it offers quantitative information on the output noise and harmonic distortion. Since the output of ADCs is discrete in time, discrete Fourier transform (DFT) is required to produce the frequency spectrum. For a discrete-time signal sampled every T_{CK} seconds, $x(nT_{CK})$, $n = 0, \ldots, N - 1$, the DFT is defined as

$$X(m) = \sum_{n=0}^{N-1} x(nT_{CK})e^{-2j\pi nm/N}. \tag{9.7}$$

Since the total number of arithmetic operations in (9.7) is proportional to N^2, more efficient algorithms (fast Fourier transforms) have been devised to make this number proportional to $N \log_2 N$. Such algorithms typically require that N be a power of 2.

The discrete nature and finite length of the data and the transform give rise to several issues [5]. First, because the frequency resolution of FFT is inversely proportional to N, a large number of samples must be taken if a narrow band of the spectrum is to be examined closely. Second, since the DFT assumes the data record is periodic, the first and last samples must exhibit no discontinuity; i.e., the record length must be an integral number of input cycles. This can be accomplished by locking the input and clock signals to the same time base such that, as explained before, the relationship $nT_{in} = mT_{CK}$ is guaranteed.

If the input and clock frequencies are independent, then any discontinuity at two ends of the data record leads to "leakage" in the frequency domain. This phenomenon occurs because, as depicted in Figure 9.18, the time domain input is multiplied by a unity-amplitude rectangle function, causing their spectra to be convolved in the frequency domain. Since the spectrum of a rectangle function (the sinc function) has finite "side lobes," parts of the higher-frequency bands leak into the base band, thus corrupting the FFT characterization. This effect can be suppressed by replacing the rectangle function with other "windowing" functions that have smaller side lobes. Several windowing functions along with their trade-offs are described by Harris [9] and Nuttall [10].

In an FFT test near the Nyquist rate, the harmonic components are aliased back into the range *below* the input frequency. Thus, it is important to select the input and sampling frequencies so that the aliased components do not coincide with the fundamental.

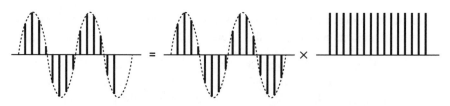

Fig. 9.18   Effect of discontinuities at the ends of data record.

Despite these difficulties, the FFT test is extensively used to characterize ADCs because its errors can be reliably controlled and reduced. Furthermore, it requires many fewer points than the histogram test and is therefore more efficient.

Sine Fitting Test.   The sine fitting test is another approach to calculating the overall signal-to-noise ratio and signal-to-(noise + distortion) ratio of ADCs. In this test, the output samples corresponding to a sinusoidal input are collected and a curve fitting program is used to determine a sine wave that fits the data with minimum rms error (Figure 9.19). This error is minimized by adjusting the fit parameters, frequency, phase, amplitude, and offset [5]. The difference between the collected data and the fitted sine thus provides quantitative information on the noise and harmonic distortion contents.

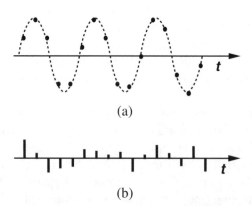

(a)

(b)

Fig. 9.19   (a) Fitting a sine to output samples of an ADC; (b) the fitting error plot.

The sine fitting test can be utilized to study the effect of white noise, jitter, and harmonics on the output SNDR. Since contribution of white noise does not depend on frequency or input amplitude, the rms error term becomes almost constant when white noise is dominant. The effect of jitter can be observed by varying the input slew rate, i.e., input amplitude or frequency, and examining the rms error at zero crossings of the sine wave. As for harmonics, the *difference* between the actual waveform and the fitted sine wave can be considered an error term that includes all the higher harmonic components. Thus, if sine waves with higher harmonic frequencies are fitted to this difference, harmonic distortion can be estimated [5].

The curve fitting program usually requires an initial guess for the fit parameters and may also have convergence difficulties if the data quality is very poor [5] or the number of points insufficient.

REFERENCES

[1] *The Handbook of Linear IC Applications*, Burr-Brown Corp., Tucson, Arizona, 1987.

[2] K. Poulton, J. S. Kang, and J. J. Corcoran, "A 2 GS/s Sample and Hold," *Proc. GaAs IC Symp.*, pp. 199-202, 1988.

[3] H. Hikawa and S. Mori, "A Digital Frequency Synthesizer with a Phase Accumulator," *Proc. IEEE ISCAS*, pp. 373-376, 1988.

[4] W. Kester, "Test Video A/D Converters under Dynamic Conditions," *EDN*, pp. 103-112, Aug. 18, 1982.

[5] B. E. Peetz, A. S. Muto, and J. M. Neil, "Measuring Waveform Recorder Performance," *Hewlett-Packard J.*, pp. 21-29, Nov. 1982.

[6] *Dynamic Performance Testing of A to D Converters*, Hewlett-Packard Co., Product Note 5180A-2.

[7] J. Doernberg, H. S. Lee, and D. A. Hodges, "Full-Speed Testing of A/D Converters," *IEEE J. Solid-State Circuits*, vol. SC-19, pp. 820-827, Dec. 1984.

[8] B. Ginetti and P. Jespers, "Reliability of Code Density Test for High Resolution ADCs," *Electron. Lett.*, vol. 27, pp. 2231-2233, Nov. 21, 1991.

[9] F. G. Harris, "On the Use of Windows for Harmonic Analysis with the Discrete Fourier Transform," *Proc. IEEE*, vol. 66, pp. 51-83, Jan. 1978.

[10] A. Nuttall, "Some Windows with Very Good Sidelobe Behavior," *IEEE Trans. Acoustics, Speech, Sig. Process.*, vol. ASSP-29, pp. 84-91, Feb. 1981.

Index